Building Math Models in Excel

Building Mathematical Models in Excel

A Guide for Agriculturists

Christopher Teh Boon Sung

Universal-Publishers
Boca Raton

Building Mathematical Models in Excel: A Guide for Agriculturists

Universal-Publishers
Boca Raton, Florida • USA
2015

ISBN-10: 1-62734-038-6
ISBN-13: 978-1-62734-038-0

www.universal-publishers.com

Cover image © Lexmomot | Dreamstime.com

Microsoft® Excel® are registered trademarks of
Microsoft Corporation.

Contents

PART II
BUILDING A CROP GROWTH MODEL

Contents

PART III
RELATIONSHIPS BETWEEN VARIABLES AND FORMULAS

CHAPTER 11. CELLS NETWORK MAP   283

REFERENCES   301

APPENDIX A. INSTALLATION OF BUILDIT   309

APPENDIX B. BUILDIT MENU COMMANDS   319

INDEX   323

Dedicated to my son, Zachary

Preface

A model is a simplified representation of a real system. A model can be a picture, illustration, verbal or text description, or even a physical replica (mockup) of the real system.

However, one type of model that is of particular interest to scientists is the mathematical model because it represents a real system using mathematical principles. The properties and behavior of the real system are represented as one or, more typically, a series of equations which can be used to further understand the real system or to make predictions about the real system in certain scenarios or conditions.

Developing a mathematical model requires three steps. The first step is to uncover and understand the science behind the real system. The second is to translate all the essential properties and behavior of the real system into a meaningful and representative mathematical form. And the third step is to implement this mathematical form into a computer program that can be executed by computers.

So, unless we are dealing with very simple equations, computers are an essential tool in mathematical modeling. A mathematical model often consists of many equations, some of which too complex, tedious, or complicated to be reliably calculated by hand. Moreover, a mathematical model often involves iterative calculations where the same set of equations are repeatedly used in calculations.

This means mathematical modeling requires agriculturists to not only be knowledgeable in agriculture science but also to be proficient in mathematics and in computer programming.

Unfortunately, most agriculturists receive no to very little formal training in computer programming.

Consequently, they often struggle with this last step of mathematical modeling; that is, to translate their models into computer program that can be correctly understood and executed by a computer. For them, computer programming often becomes a tedious and time-consuming drudgery that distracts them from their main purpose of study or work.

Consequently, this book is targeted at agriculturists particularly those who are either novices or non-programmers and who wish to have an easier way to implement their models in computers. This book shows how model implementation can be easily carried out in a spreadsheet, with the spreadsheet of choice being Microsoft Excel.

But why spreadsheets, and why Excel? This is because spreadsheets provide a modeling platform that requires the least proficiency in computer programming. Unlike other modeling platforms that enforce a rigid programming structure, spreadsheets' unrestricted and open structure enable novices and non-programmers to easily implement their models in a spreadsheet and to have the spreadsheet run the model simulations. One spreadsheet program in particular is Excel. It is popular and the most widely used spreadsheet in the world – for several reasons. Excel is easy to use, versatile, and powerful, and it has a very gentle learning curve. Even first-time Excel users can quickly learn how to enter their data in Excel and have Excel perform calculations on them using formulas and functions.

Nonetheless, Excel does have several key limitations that prevent the implementation of large, complex models. To circumvent these limitations, Excel's programming language, VBA (Visual Basic for Applications), could of course be used. However, using VBA requires programming skills in which most agriculturists lack. Moreover, requiring agriculturists to learn VBA would defeat the purpose of having an alternative and easier way for agriculturists to implement their models.

To overcome Excel's limitations, an Excel add-in, called *BuildIt*, was developed (Teh, 2011). Although BuildIt was developed using VBA, BuildIt shields users from VBA. With BuildIt, users are able to implement simple to complex models in Excel without requiring knowledge in VBA or a strong proficiency in computer programming.

BuildIt removes—or at least, greatly reduces—the distraction of computer programming and allows agriculturists to concentrate on the more important task of building their mathematical models and using them in their studies or work.

This book is a guide for agriculturists to learn how they can implement their mathematical models in Excel with the support from BuildIt. This book uses examples related to agriculture, but there is no reason why Excel or BuildIt cannot be used to implement models from other disciplines.

Examples in this book start by using simple models, but they soon include progressively larger, more complex, and more realistic agriculture models. By the end of the book, agriculturists would have learned how to implement a generic crop growth model that is made up of five components: meteorology, photosynthesis, energy balance, soil water, and crop growth development. Exercises included at the end of chapters are to test understanding and to enable the models to be run in other conditions.

Chapter 1 in this book starts by discussing about the different types of modeling platforms. This chapter focuses primarily on why using spreadsheets like Excel can be a suitable modeling platform for novices or non-programmers. However, the weaknesses and limitations of spreadsheets as a modeling platform are also discussed. Lastly, good programming practices are listed at the end of this chapter

Part I of this book encompasses Chapters 2 to 5, and these chapters cover what BuildIt is and how it can be used to implement mathematical models in Excel. Several examples of models are used. Simple, rather than large, complex models, are used as examples so that certain key aspects in model implementation can be highlighted without the distraction of excessive details.

Chapters 3 and 4, in particular, discuss about BuildIt *actions*. These actions circumvent the limitations of Excel that would normally hinder the implementation of large, complex models. These actions, for instance, allow a spreadsheet cell to manipulate or change the content of another cell, or for a cell to update its own value. Both these operations are conventionally not allowed in Excel. In Excel, all external cell are strictly read-only. Cell A1, for instance, can read but cannot alter the content in cell B1. Likewise, cell A1 cannot update its own value, say, by having the following formula "$=A1+1$", where cell A1 refers to its current value, increments it by 1, then updates itself to have this new value. This self-referencing operation is not allowed in Excel because it would result in a never-ending loop of recalculations in cell A1. BuildIt actions remove such restrictions imposed by Excel and without causing the negative consequences. Typical mathematical operations like integrations and differentiations that are often needed in mathematical models are also supported by the BuildIt actions.

Chapter 5 discusses about scenarios. BuildIt setups a system or structure that allows a single model run to execute two or more scenarios in a sequential manner. BuildIt also provides three custom macros and two functions. Two of these macros deal with freezing and unfreezing the screen updates. By freezing the screen updates, the model run can be speeded up because Excel does not have to always refresh or update the screen whilst the model is running. The last macro deals with deleting the model output from the previous model run. BuildIt supplies two custom functions, `interpolate` and `solve`. The function `interpolate` is used for linear interpolation between two given values, and `solve` is to solve simultaneous linear and nonlinear equations.

Part II of the book encompasses Chapters 6 to 10. These chapters show how a non-trivial generic crop growth model is built sequentially from its five model components, starting from the meteorology component (Chapter 6), photosynthesis component (Chapter 7), energy balance component (Chapter 8), soil water component (Chapter 9), and finally, the crop growth development component (Chapter 10).

The final chapter, Chapter 11, shows how a tool of BuildIt, called *Trace*, can be used to draw a visual map of how spreadsheet cells are related to one another. By examining the cells network map, agriculturists can better understand how variables and formulas are related to one another. This can be particularly useful to detect errors in a model or to better understand how a model works.

At the end, the lessons learned from this book would allow agriculturists, as well as non-agriculturists, to implement their own mathematical models in Excel.

It is important to note that the intention of developing BuildIt was to overcome some of Excel's weaknesses so that simple and complex models can be implemented in Excel. The purpose of BuildIt is not to substitute or compete against other modeling platforms, but to be a useful tool that novices and non-programmers can rely on in their modeling work.

BuildIt was first published in the Journal of Natural Resources & Life Sciences Education (Teh, 2011).

Christopher Teh Boon Sung
Serdang, Malaysia
March 3, 2015

Chapter 1. Spreadsheets for mathematical modeling

1.1 Types of modeling software (platform)

Mathematical modeling often incurs a steep learning curve to many agriculturists. This is partly because one of the stages in the modeling process involves computer programming, a skill in which many agriculturists lack. Consequently, many agriculturists struggle to translate their formulated models into a set of instructions that can be implemented by a computer system. For them, computer programming often degenerates into a tedious and distracting undertaking.

This problem has been recognized even as early as four decades ago when Hillel (1977) commented that agriculturists, being novice programmers, are easily entangled in the complexities of translating their model into a computer source code. Rather than focusing on the important—and intended—task of mathematical modeling and simulation, they instead become distracted by the intricacies and mechanism of computer programming.

What is required then is some special software that allows non-programmers to build their own computer models. This will enable non-programmers, agriculturists included, to accomplish their modeling tasks directly and quickly and without the distraction of having to learn a new computer language (or to master a new simulation application).

There are generally four categories of software platforms to aid in model building and simulation. They are: 1) general purpose computer languages (*e.g.*, C, C++, FORTRAN, BASIC, Pascal and Java); 2) specialized

1

simulation applications (*e.g.*, FST, PowerSim, Stella and ExtendSim); 3) equation solver-based applications (*e.g.*, Maple, Mathematica, and Mathlab); and 4) spreadsheet-based applications (*e.g.*, Microsoft Excel and OpenOffice Calc)[1]. Of these four groups of software platforms, spreadsheet applications require the least level of proficiency in computer programming.

It is no surprise then that spreadsheets have been advocated by several workers (*e.g.*, Seila, 2005; Brown, 1999; Nardi and Miller, 1990) as a suitable platform, especially for non-programmers, to build and simulate their models.

Spreadsheets are immensely popular and widely used because they are easy to use and versatile. Additionally, they provide the following features (adapted from Seila, 2005): 1) a large number of numerical and non-numerical functions (*i.e.*, dedicated formulas) to do mathematical, statistical, database, date and time, financial, engineering, and other types of calculations; 2) database representations and access; 3) charting and graphing; 4) display and document formatting capabilities such as layout, fonts, and colors to improve presentation; and 5) scripting or

---

[1] FST (Fortran Simulation Translator) is available upon request from its developers (*see* van Kraalingen *et al.*, 2003);
Powersim® is a registered trademark of Powersim Software AS;
Stella® is a registered trademark of isee systems;
ExtendSim® is a registered trademark of Imagine That Inc.;
Maple® is a registered trademark of Waterloo Maple Inc.;
Mathematica® is a registered trademark of Wolfram Research Inc.;
Mathlab® is a registered trademark of The MathWorks Inc.;
Microsoft® Excel® are registered trademarks of Microsoft Corporation;
OpenOffice® is a registered trademark of The Apache Software Foundation.

programming language such as VBA (Visual Basic for Applications) in Excel and OpenOffice Basic in OpenOffice Calc.

Spreadsheets were initially conceived as electronic accounting books for financial analysis, but today, spreadsheets have progressed beyond their original intent. They have become powerful tools to manipulate, analyze, and present data, and to build models for simulating various phenomena in science and engineering (Khandan, 2001).

1.2 Advantages and disadvantages of spreadsheets

Spreadsheets are well known that they are easy to use, versatile, and powerful, but it is not so much these characteristics that qualify spreadsheets as a suitable modeling platform for non-programmers.

The first key benefit of using spreadsheets in modeling is that they do not require users to be proficient in computer programming. This is achieved by the spreadsheet system providing: 1) automatic control and maintenance of the program flow, 2) a simple, straightforward modeling framework, and 3) high-level and task-specific functions (Nardi and Miller, 1990).

A spreadsheet is a group of pages, or worksheets, where each worksheet is a two-dimensional table consisting of rows and columns of cells. Each cell can contain either a constant or a calculated value, derived from a function or formula. The open tabular format of the spreadsheet provides users a simple, straightforward framework in which to build their models.

As such, spreadsheet users only need to understand two basic concepts: to treat the cells as variables and the functions (or formulas) as the relationship between these variables. Users specify the way variables depend on each other via formulas or functions, and the spreadsheet system maintains these variable dependencies. So, if one part of the spreadsheet changes, it triggers an update to the whole spreadsheet so that all the dependent variables are automatically recalculated to reflect their new values— giving users immediate feedback. The order in which the variables are calculated is worked out by the spreadsheet system based on the variable dependencies. Users cannot contravene this calculation sequence by giving the spreadsheet system a different calculation or action sequence to be performed. Though this might seem restrictive, it is actually this feature, among others, that makes spreadsheets appealing to non-programmers.

Spreadsheets not only relieve users from having to maintain the program flow themselves, but also from having to design their own modeling framework. Flow control and framework design are both difficult programming concepts for non-programmers to master (Nardi and Miller, 1990; Hoc, 1989; Lewis and Olson, 1987; Soloway et al., 1983). Without automatic flow control, users will have to write their own programming code to track variable dependencies and to update all effected variables iteratively when required. And users need not design their own modeling framework because the spreadsheet already provides users with one.

Spreadsheets further cater to non-programmers by providing high-level and task-specific functions, meaning that these functions can be used without users having to understand how they work or to require computer programming expertise. One commonly used function in Excel, for example, is the SUM function that, as its names implies, calculates the total of a set of given values.

4

Spreadsheet users merely write the SUM function and specify all the cells that contain the values to be added. Without such a high-level function, users will require to write a loop to iterate through the elements in an array, summing each element and accumulating the total. Other routinely used functions are such as AVERAGE (for calculating the mean of values), MIN and MAX (for determining the smallest and largest value, respectively, from a given set of values), VLOOKUP and HLOOKUP (for looking up a value from a vertically and horizontally tabulated data, respectively), IF (for conditional computations), and trigonometric functions like SIN and COS. These are only a few of the functions listed; hundreds more functions are available in Excel (as well as in other spreadsheets) to support users' requirements.

The second key benefit of using spreadsheets is its table-oriented interface. It provides a strong visual format to manipulate, organize and present data as well as a problem-solving tool (Nardi and Miller, 1990). The spreadsheet's open tabular format allows direct access to the variables' values. This is in contrast to traditional programming efforts which require some special and intentionally constructed mechanisms for the input and output of variables. Brown (1999) additionally noted that because the values of all variables are always updated and displayed in the table, this could help to detect the location of errors during model construction by the presence of nonsensical or questionable values.

The spreadsheet's tabular format additionally allows users to organize visually their model into subparts or modules, where each module performs a specific task to solve an overall, larger problem. This organization can be done by segmenting the large model into smaller modules by, for instance, leaving an empty row between each module or by placing each module in a separate worksheet. Formatting the module sections such as by applying different fonts, shading, style layout or colors further help in code organization. Consequently, the model's code is distributed visually over one or more grids in such a way that makes the model more comprehensible and maintainable. Visual code organization is not merely for aesthetic reasons. The very act of laying out data, organizing and formatting the code provide an important visual feedback mechanism that helps to organize users' thoughts and shape their problem solving process (Nardi and Miller, 1990).

Spreadsheets, however, have several important limitations. Spreadsheet models can be difficult to understand when studied by others. As mentioned previously, the modeling framework provided by spreadsheets is a simple, straightforward structure, essentially consisting of variables (cells) and their linkages with each other (formulas). But as models increase in complexity, these variable linkages can grow into an intricate network of relationships. Since the spreadsheet system maintains the program flow and shields users from viewing this network of relationships, it can be difficult for other users (even for the model developers themselves) to grasp the computations as a whole. This causes difficulty not only in understanding the model but also in debugging it (that is, to locate and correct the errors in the model).

Spreadsheets offer to non-programmers a modeling platform that is almost unrestrictive and freeform and with little validation checks. While this feature can be convenient for those who wish to develop their models easily and quickly, it does, however, make it easy for users to introduce errors unwittingly into their models. Errors in large spreadsheet models, in particular, can be difficult to locate and, at times, be undetected.

In the business sector, the frequency of errors in their spreadsheets are reported to be alarmingly high. Rajalingham et al. (2001) reported that as much as 90% of real-world spreadsheets contained errors. Furthermore, the European Spreadsheet Risks Interest Group (www.eusprig.org) lists nearly a hundred examples of newsworthy incidences where spreadsheets errors caused companies either financial losses, incorrect forecasting, or simply, embarrassment. Although spreadsheets from the business sector are often cited as examples of high occurrence of errors, we cannot disregard the possibility that spreadsheets from other disciplines (such as agriculture) are plagued too by high number of errors.

The simple modeling framework provided by the spreadsheet also makes spreadsheet models difficult to *reuse* and *extend*. Reusability and extendibility are two software engineering concepts related to ways to increase the usefulness, applicability and lifespan of software. A modeling framework that adheres to these two concepts means that, ideally, the framework is flexible and adaptable enough for algorithm modification, substitution and addition. In some ways, reusability and extendibility are analogous to the "plug-and-play" feature in modern computer systems. For instance, adding, replacing or removing a computer peripheral (such as a printer, scanner or mouse) should be a seamless operation that does not disrupt a computer system or cause it to malfunction. In a similar way, a reusable and extendible model means that one or more parts in the model can be modified easily or even substituted with other parts from another model. It also means that additional features or functions can be added to the model. All these operations can occur without having to break or redesign the existing modeling framework.

There are several examples in agriculture of modeling frameworks developed with concerns of achieving reusability and extendibility (*e.g.*, Papajorgji and Pardalos, 2006; Papajorgji et al., 2004; Hillyer et al., 2003; Caldwell and Fernandez, 1998; Acock and Reddy, 1997).

Nevertheless, until today, none of them has yet to be adopted widely by agriculturists. One could speculate that these framework designs are still not sufficiently adaptable and flexible enough to cope with the unpredictable and transient requirements of agricultural models. Or it may simply be that agriculturists do not have the interest or time to learn about reusability and extendibility, both concepts of software engineering, not agriculture. Consequently, for many agriculturists, the problem that spreadsheet models are difficult to reuse and extend may actually be an unimportant concern in their work.

Spreadsheets have other limitations that are particularly relevant for models with complex data structures and algorithms (Seila, 2005).

A spreadsheet is a two-dimensional table of cells, and while this format is adequate in most cases, it is a limitation when more elaborate data structures are required such as lists and trees. Although spreadsheets can be coerced to handle such complex data structures, the way they are achieved in spreadsheets can be inefficient and convoluted. Complex algorithms can be particularly difficult to implement in a spreadsheet.

Spreadsheets lack explicit loop and conditional controls commonly found in general purpose languages.

Spreadsheets actually have a conditional construct IF-THEN-ELSE, but its effect is only local to the cell that contains the construct. The construct cannot be used to change the values in other cells or to transfer control to another cell or to another part of the spreadsheet. As stated earlier, spreadsheets relieve users from having to maintain the control flow themselves. While this is a convenient feature for non-programmers, it becomes a severe limitation when an algorithm requires users to specify explicitly the sequence of actions to be performed.

Spreadsheets like Excel and OpenOffice.org Calc have their own scripting language that can be used to circumvent this problem, but they require proficiency in computer programming. But non-programmers may find this solution unappealing because their use of spreadsheets is precisely to avoid computer programming in the first place.

Lastly, spreadsheet models often run slower than models developed in other software platforms. This is partly because models developed in other platforms, such as in C, C++, or FORTRAN, are compiled (translated) into fast machine code that is in a format directly executable by the computer.

In contrast, no such compilation occurs for spreadsheet models. The formulas in spreadsheets are instead interpreted first before execution. This slows down program execution. Dalton (2005) roughly estimated that models written in Excel could be slower by as much as five to twenty times than if they were written in C and C++ computer languages. He further reported that Excel's text-manipulation routines are very much slower than the same routines in C and C++ by as much as over 300 times. Slow execution speeds may not be noticeable for simple

9

spreadsheet models, but for larger, more complex models (due to their more numerous and intricate calculations), their execution speeds can be noticeably sluggish and become a significant constraint.

For non-programmers, spreadsheets offer an infrastructure that appeals to them because spreadsheets allow them to build fully functioning models quickly and directly, and without the pre-requisite knowledge in any advance programming concepts and techniques. But for those with programming expertise, the limitations of spreadsheets can be significant reasons why spreadsheets are not used as their modeling platform. In some cases, spreadsheets are merely used as a quick test bed to test their modeling ideas or to develop model prototypes. Once these ideas or model prototypes have been assessed sufficiently, model development then shifts to other software platforms for completion.

1.3   Guidelines to programming in spreadsheets

Scoville (1994) remarked, "Spreadsheets make it easy to do complex calculations – and even easier to do them incorrectly."

Spreadsheet's lack of modularity, structure, and data-checking unfortunately encourage bad programming practices which not only produce spreadsheet models that are difficult to understand and maintain but also make it easy to unwittingly introduce errors into the models.

Consequently, the following are guidelines, adapted from Kruck (2006), Raffensperger (2003), and Read and Batson (1999) that should be adhered when using spreadsheets for developing our models:

a) Have the spreadsheet read from left to right and top to bottom. Just as we read from the left to right and from top to bottom, the flow of information in a spreadsheet should likewise follow the same. Formulas in cells should depend on cells only above and to the left. An example of violating this guideline is when a formula in cell B6 requires information from cell F10 (which is to the right and below cell B6), which in turn requires information from cells A1 and H10. This kind of layout causes information to flow "back and forth" in a seemingly disorganized or haphazard manner.

b) Minimize the distance between cell dependents. Cells which are logically related and depend on each other for information should be placed as closely together as possible, preferably in the same row or column. When closely-related cells are placed together, the spreadsheet organizes itself into blocks of common relationship, meaning, or purpose.

c) Split the implementation of a model into two or more worksheets, so that each worksheet implements a small but specific component of the model.

A crop growth model, for instance, typically has meteorological, photosynthesis, respiration, energy balance, and water balance components. Each of these five model components are implemented as separate worksheets of the same spreadsheet. An additional two worksheets should also be created: one for model data (*i.e.*, inputs; that is, where we input or give values to model parameters) and the second for model output (where the results of simulations are stored).

Even for very simple models where it may not be necessary to have multiple worksheets, we still need to organize the spreadsheet layout so that the model input, computations, and output areas are kept separated from one other in distinct areas of a worksheet.

d) Split complex formulas. Like breaking up a large model into two or more components, a complex or long formula should also be broken up into two or more parts. Avoid creating a "mega" equation where the full equation is implemented within one spreadsheet cell.

e) Use short and descriptive labels to identify variables, formulas, and operations. Labels provide documentation to users which help to make code more easily understandable.

f) Formulas should only contain cell references for equation parameters, even if the equation parameters are constants. The value for an equation parameter should be placed in a separate cell, and all dependent formulas should instead refer to the parameter's cell address. When a constant is hard-coded into formulas, users must remember to change all the relevant formulas if a different constant value is needed. Failure to change all the relevant formulas will introduce subtle errors in the model because one or more formulas are still using the previous constant value.

For instance, instead of : "=98*B10", where the value 98 is hardcoded into the formula, we ought to use "=A2*B10", where cell A2 contains the value 98. When the value in cell A2 is changed, all formulas that refer to cell A2 will recalculate to reflect their new values. This avoids having us to remember or track down all formulas that use the "old" value of 98.

An exception to this guideline is if an equation contains parameters that are either immutable (*e.g.*, physical constants like the gravitational constant, Stefan-Boltzman constant, and gas constant) or very unlikely to change.

g) Erase isolated cells. Examples of isolated cells are those that contain unused calculations, input, or data. Since these cells are unused, they obfuscate the code, and they should be deleted.

h) Use cell names instead of cell addresses for model parameters and commonly used variables. Descriptive cell names improve formula readability and reduce typing error in formulas.

A formula that uses cell names such as "=Am*e*k*Io*EXP(-k*L)" is more meaningful than its equivalent that directly uses cell references such as "=B2*B3*B4*E5*EXP(-B4*E6)".

PART I
Using BuildIt

Chapter 2. Building mathematical models in Excel

2.1 Introducing BuildIt

BuildIt is an add-in that works within Excel to support model building and simulation in the spreadsheet. Excel is the spreadsheet of choice because Excel is by far the most widely used spreadsheet in the world today, with an estimate of 400 million users worldwide and preferred by more than 70% of spreadsheet users (Syrstad and Jelen, 2004; Liebowitz and Margolis, 1999).

BuildIt is written in Excel VBA (Visual Basic for Applications) which is Excel's programming language. However, you do not need to understand or use Excel VBA to build your models in Excel. BuildIt shields you from having to learn Excel VBA.

It is important to understand that the primary purpose of developing BuildIt is neither to supplant other software modelling platforms nor to compare BuildIt's features with other modelling platforms. BuildIt is also neither a mathematical model nor a "toolbox" of useful equations.

Instead, the primary purpose of BuildIt is to overcome some of Excel's limitations so that Excel could be used more effectively to develop and test mathematical models, as well as to run model simulations. The challenge is to develop a tool that makes Excel a modelling platform but without requiring users to program in Excel VBA. BuildIt allows non-programmers to build their models, especially those models that are large and complex and those that have iterative (repetitive) calculations.

Crop models, for instance, are often large and complex because they comprise several interrelated components

Example 1. A simple quadratic equation

such as meteorology, water and energy balance, and plant photosynthesis and respiration. Crop models often also involve iterative calculations, where the same set of calculations are repeated many times (*e.g.*, at daily time steps) to simulate the dynamics of plant growth and its environmental conditions. To develop such complex models in native Excel is difficult, but with BuildIt, such models could be built entirely within the spreadsheet.

The best way to quickly understand how BuildIt works and how to use it for model building is through the following examples.

But before we begin, please refer to Appendix A on the download and installation instructions of BuildIt. Rather than a lengthy preamble on covering the basic topics of how to use Excel, this book needs to assume that you are at least a fairly competent user of Excel (but knowledge in Excel VBA is not needed). Excel is fortunately easy to use, and many textbooks (as well as tutorial websites) are available for users to learn how to use Excel.

Let us now build our first mathematical model in Excel.

2.2   Example 1. A simple quadratic equation

Consider the following model which is actually just a simple quadratic equation:

$$y = ax^2 + bx + c \qquad (2.1)$$

where *a*, *b*, and *c* are the equation parameters given the values of -3, 30, and 10, respectively.

We want to use Excel to calculate the values of y for $x =$ 0, 1, 2, ..., 10. The following two sections describe how Eq. 2.1 can be implemented in Excel: first without BuildIt then with BuildIt.

2.2.1   Building the model without BuildIt

Fig. 2.1 depicts one way Eq. 2.1 can be implemented without using BuildIt in an Excel worksheet. Cells B2, B3, and B4 contain the values for the parameters *a*, *b*, and *c*, respectively; cells A7 to A17 (A7:A17) contain the values for *x* from 0 to 10; and each of the formulas in the cell range B7:B17 implement Eq. 2.1 for *x* = 0, 1, 2, ..., 10, respectively.

	A	B
1	**Parameters**	
2	a	-3
3	b	30
4	c	10
5		
6	**x**	**y**
7	0	=$B$2*A7^2+$B$3*A7+$B$4
8	1	=$B$2*A8^2+$B$3*A8+$B$4
9	2	=$B$2*A9^2+$B$3*A9+$B$4
10	3	=$B$2*A10^2+$B$3*A10+$B$4
11	4	=$B$2*A11^2+$B$3*A11+$B$4
12	5	=$B$2*A12^2+$B$3*A12+$B$4
13	6	=$B$2*A13^2+$B$3*A13+$B$4
14	7	=$B$2*A14^2+$B$3*A14+$B$4
15	8	=$B$2*A15^2+$B$3*A15+$B$4
16	9	=$B$2*A16^2+$B$3*A16+$B$4
17	10	=$B$2*A17^2+$B$3*A17+$B$4

Fig. 2.1. The quadratic equation $y = ax^2 + bx + c$ implemented without BuildIt in an Excel worksheet.

Note that, following the good programming spreadsheet practices (*see* section 1.3), the values for *a*, *b*, and *c* are not hardcoded in the formulas. Instead, their values are referred

Example 1. A simple quadratic equation

to other cells: cell B2 for *a*, B3 for *b*, and B4 for *c*. And because *a*, *b*, and *c* do not change (*i.e.*, are constants), their cell references should be absolute. However, *x* varies between 0 to 10, so cell references to them (A7 to A17) should be relative.

In cell B7, for example, the formula is:

$$= \$B\$2 * A7 \wedge 2 + \$B\$3 * A7 + \$B\$4$$

where \$B\$2, \$B\$3, and \$B\$4 are the absolute cell references to the values for *a*, *b*, and *c*, respectively, and A7 is the relative cell reference which contains the value 0 (*x*=0). Similarly, in cell B8, the formula again uses cell references \$B\$2, \$B\$3, and \$B\$4 for *a*, *b*, and *c*, respectively, but cell reference A8 for *x*=1.

Fig. 2.2 shows the model results, and the function plot of Eq. 2.1 for *x*=0, 1, 2, ..., 10 is additionally drawn.

	A	B	C	D	E	F	G
1	Parameters						
2	a	-3					
3	b	30					
4	c	10					
5							
6	x	y					
7	0	10					
8	1	37					
9	2	58					
10	3	73					
11	4	82					
12	5	85					
13	6	82					
14	7	73					
15	8	58					
16	9	37					
17	10	10					

Fig. 2.2. Results of $y = ax^2 + bx + c$ for x = 0, 1, 2, ..., 10 in Excel (without using BuildIt)

2.2.2 Building the model with BuildIt

Typically, there are three sections we need to setup whenever we use BuildIt to implement our models: 1) the loop or control section, 2) the what-to-output and output sections, and 3) the calculations section.

The important section is the loop or control which contains information used by BuildIt so that calculations are carried out repeatedly until a certain condition is met to end the calculation loop.

The what-to-output and output sections are related to each other. The what-to-output section tells BuildIt what model results we want to be saved or displayed. For a crop model, for instance, we typically want the crop's growth parameters such as plant weight, plant height, and leaf area to be displayed or stored at daily time steps. Once we have instructed what model results we want BuildIt to store, we need to setup the output section. The output section tells BuildIt where we wish our output to be stored in the Excel workbook. In other words, the what-to-output and output sections are the "*what*" and "*where*" of the model results: the what-to-output section tells BuildIt *what* model results we wish to see and the output section *where* our selected model results are to be placed.

Lastly, the calculations section is the heart of the model implementation because this section comprises all the equations (implemented as Excel formulas) of the model.

Fig. 2.3 shows how Eq. 2.1 is implemented using BuildIt in an Excel worksheet.

Cell range B2:B4 holds the values for the parameters a, b, and c.

Cell range B7:B10 is the loop or control section which contains the information needed for BuildIt to maintain the loop.

Example 1. A simple quadratic equation

Cell range D8:E8 is the what-to-output section and D13:E13 the output section.

The cell range E2:E3 is the calculations section.

	A	B	C	D	E	F
1	Parameters			MODEL		
2	a	-3		x	=_step	
3	b	30		y	=$B$2*E2^2+$B$3*E2+$B$4	
4	c	10				
5						
6	CONTROL			TO OUTPUT		
7	maxsteps	10		x	y	
8	stepsize	1		=E2	=E3	
9	step					
10	criteria	=E2<=B7				
11				OUTPUT		
12				x	y	
13						
14						

Fig. 2.3. The quadratic equation $y = ax^2 + bx + c$ implemented in an Excel worksheet using BuildIt.

Text labels (such as those in cell ranges A1:A10, D1:D3, D7:D8, and D12:E12) are used to briefly document or describe the various values and equations. These text labels are not necessary for running the model, but they help to document the model and make understanding the model implementation easier.

Let us now look at the individual sections.

2.2.2.1 Loop section

BuildIt's loop mechanism allows for iterative or repetitive calculations to occur in Excel.

In our case, Eq. 2.1 can be used repeatedly, first to determine y for $x = 0$, then $x = 1$, $x = 2$, and so on until $x = 10$.

The control section (cell range B7:B10) in Fig. 2.3 contains the information BuildIt needs to maintain the loop (Fig. 2.4).

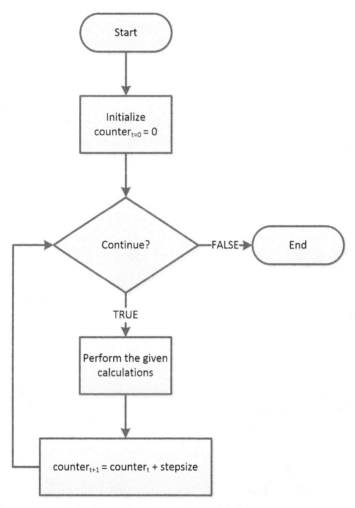

Fig. 2.4. BuildIt loop for iterative calculations

Every BuildIt loop contains only one counter, and this counter holds a value which is incremented by a given interval or step size at the end of every loop cycle:

Example 1. A simple quadratic equation

$$counter_{t+1} = counter_t + stepsize \qquad (2.2)$$

where *counter_t* and *counter_{t+1}* are the counter values at iteration step t and $t+1$, respectively, and *stepsize* is the size by which the loop counter will be increased at the end of every loop cycle. A loop must also contain information about when the loop should end its run; otherwise, the loop will continue indefinitely (*i.e.*, a never ending loop).

Before the loop cycle begins, the *counter* is initialized by BuildIt with the value 0 (Fig. 2.4).

The loop condition is next checked to determine if the first loop cycle should begin. If the criteria to start the loop are satisfied, the loop starts by performing the given calculations (*e.g.*, Eq. 2.1).

After the calculations, the *counter* is incremented by the interval size *stepsize* (Eq. 2.2). After the *counter*'s value is updated, the loop condition is checked again to determine if the loop should continue with the next cycle. If the criteria to continue are met, the next loop cycle begins; otherwise, the loop ends, ending all calculations too.

In Fig. 2.3, cell B9 holds the loop counter, and regardless what value we enter in cell B9, BuildIt will always initialize this counter with the value 0 at the start of every model run (Fig. 2.4). Cell B8 holds the interval or step size and cell B10 the criteria to resume or end the loop run.

But how would BuildIt know that cells B8, B9, and B10 hold the loop's step size, counter, and condition, respectively?

To instruct BuildIt on the location of these three loop parameters, we need to define cells with the following cell names: _stepsize for the step size, _step for counter, and _criteria for the loop's condition (Table 2.1).

Table 2.1. Cells B8, B9, and B10 are defined with the BuildIt cell names `_stepsize`, `_step`, and `_criteria`, respectively, so that BuildIt knows where the information about the loop are located in the spreadsheet.

Cell	Cell name	Description
B8	`_stepsize`	Interval or step size to increase the loop counter at the end of every loop cycle
B9	`_step`	Loop counter
B10	`_criteria`	Criteria on whether to continue (if TRUE) or end (if FALSE) the loop run

Since we want cell B8 to hold the loop's step size, we define the name `_stepsize` to cell B8. Similarly, we want cell B9 to hold the loop counter, so this cell is defined with the cell name `_step`. Lastly, cell B10 is defined with `_criteria`, so that BuildIt knows where to check the logical condition if the loop should continue or end. You will notice that these BuildIt's pre-defined cell names all begin with an underscore. If one or more of these three cell names are not defined, BuildIt will emit an error message prior to starting a model run.

Cell B10, as stated earlier, holds the criteria on whether to continue or end a loop run. As long as the logical condition remains TRUE, BuildIt will continue with the loop run. A FALSE condition will end the loop. In Fig. 2.3, cell B10 contains the formula: "=E2<=B7". This formula means that BuildIt will run the loop as long as the value in cell E2 is smaller than or the same as 10 (the value in cell B7). Cell E2, as we will later see in the next section, contains the value of *x*. So, when *x* exceeds 10, the loop run ends.

Example 1. A simple quadratic equation

Why should the loop terminate when the loop counter exceeds 10? Recall that we want to determine y for $x=0$, 1, 2, and so on until 10. The last calculation occurs when x reaches 10. Once x exceeds 10, the loop and calculations must end.

2.2.2.2 Calculations section

Cell E3 implements Eq. 2.1 by the formula:

$$= \$B\$2*E2^2 + \$B\$3*E2 + \$B\$4$$

where the absolute cell references $\$B\$2$, $\$B\$3$, and $\$B\$4$ are for the parameters a, b, and c, respectively; and cell E2 holds the value for x. Note that cell E2 does not contain a numeric value but the formula "= _step". In other words, cell E2 refers to the loop counter. As stated earlier, BuildIt will always initialize the loop counter with 0 at the start of a model run. Since cell E2 refers to the loop counter, this cell will also have 0 as the first or starting value.

Starting with $x=0$, BuildIt next checks cell B10. This cell will return TRUE (since x is smaller than 10), so the calculations in cell E3 are done to return the corresponding y value for $x=0$.

BuildIt will next increment the loop counter (cell B9) by step size (cell B8) from 0 to 1. Since the loop counter is smaller than 10, the second loop cycle begins. With the new value in cell E2 ($x=1$), calculations in cell E3 are repeated to obtain the corresponding y value.

The loop counter is incremented again by 1 to 2. And since the loop counter is still less than 10, the third cycle begins. With the new value in E2 ($x=2$), the calculations in cell E3 are done again to compute the corresponding y value.

The above cycle repeats until the loop counter is incremented to 11. Since the loop counter is now larger than 10, cell B10 returns FALSE, signaling to BuildIt to end the loop.

This kind of looping mechanism allows cell E3 to iteratively recalculate to return the corresponding y value for $x = 0, 1, 2, ..., 10$.

2.2.2.3 What-to-output and output sections

The what-to-output section in the cell range D8:E8 tells BuildIt that we wish to store the values of x and y at the end of every loop cycle (Fig. 2.3). Cell D8 refers to cell E2 (which holds the value of x) and cell E8 to E3 (which holds the value of y). Now that BuildIt knows what we want to be included in the model output, BuildIt will store the x and y values as a list starting at cell D13.

We need to use two of BuildIt cell names: `_read` and `_write`, where the former tells BuildIt where to look for the what-to-output items and the latter for where to store the output items (Table 2.2).

Table 2.2. Cells defined with BuildIt cell names `_read` and `_write` tell BuildIt where to read the list of parameters to output and where to place the output, respectively.

Cell	Cell name	Description
D8	`_read`	Marks the start of the list of parameters to include in the model output
D13	`_write`	Marks the start of the model output

In Fig. 2.3, BuildIt reads the list of items to output starting from cell D8 (which has been defined with the cell

Example 1. A simple quadratic equation

name _read) in a horizontal (from left to right) manner. Cell E8 is also read since it is adjacent to cell D8. Since cell F8 is blank, BuildIt stops reading and accepts that the values of *x* and *y* are to be stored. Cell D13 should be blank because this cell has been defined with the cell name _write which marks the start of the model results listing. If cell D13 contains any value or formula, it will be overwritten by BuildIt during the model output.

Lastly, Fig. 2.5 summarizes the program flow of our model.

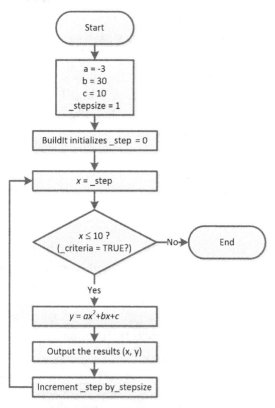

Fig. 2.5. Program flow of the model $y = ax^2 + bx + c$ developed in Excel (using BuildIt).

2.2.2.4 Model results

The model is run by clicking on the "Start Simulation" command from BuildIt menu (Fig. A.3 and A.5). Once the model run is completed, the model output will look like Fig. 2.6.

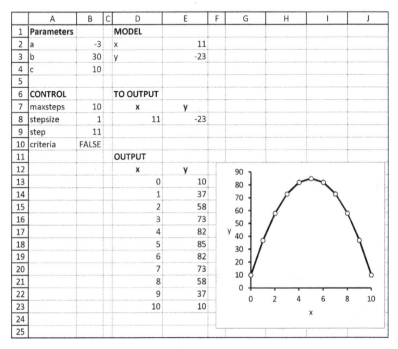

	A	B	C	D	E	F	G	H	I	J
1	Parameters			MODEL						
2	a	-3		x	11					
3	b	30		y	-23					
4	c	10								
5										
6	CONTROL			TO OUTPUT						
7	maxsteps	10		x	y					
8	stepsize	1		11	-23					
9	step	11								
10	criteria	FALSE								
11				OUTPUT						
12				x	y					
13				0	10					
14				1	37					
15				2	58					
16				3	73					
17				4	82					
18				5	85					
19				6	82					
20				7	73					
21				8	58					
22				9	37					
23				10	10					
24										
25										

Fig. 2.6. Results of $y = ax^2 + bx + c$ for $x = 0, 1, 2, \ldots, 10$ in Excel (using BuildIt)

The model output are stored as a list starting at cell range D13:E13 and ending at D23:E23. This model output (cells D13:E23) are further used to draw the chart to visually depict the model results.

Note that this chart was not drawn by BuildIt; it was separately drawn using Excel's native charting capabilities. BuildIt does not have any features for automated charting

Example 1. A simple quadratic equation

capabilities since those provided by Excel are already sufficient, powerful, and flexible.

Also note from Fig. 2.6 that the loop counter (cell B9) ends at value 11. This value is larger than 10, and this is why cell B10 returns a FALSE value, which in turn signals to BuildIt to end the loop run. Although Excel still performs the calculations to determine y for $x = 11$ (see the cells E2 and E3), their results are not stored in the model output list (see the cell range D13:E23).

Admittedly, using BuildIt to implement Eq. 2.1 in Excel appears unnecessarily complicated and less straightforward than that without using BuildIt (*e.g.*, compare Fig. 2.1 with Fig. 2.3).

However, as you will later discover, using BuildIt becomes necessary when we build large and more complex models, such as those that have a large number of equations, those that require numerous iterative calculations, or those that require calculus operations like integration.

2.2.2.5 Other scenarios for running the model

Previously, the interval or step size was set at 1, so that y was calculated using Eq. 2.1 for $x = 0, 1, 2, ..., 10$. But what if we wish to determine y for $x = 0, 2, 4, ..., 10$?

In this case, only a slight modification is required. Change the step size from 1 to 2 by entering 2 in cell B8. With this change, the loop counter will be incremented by 2, so that x will have only even numbers.

The following are some other scenarios:

a) How do we determine y for $x = 0, 1, 2, ..., 20$?
   Answer: enter 20 in cell B7 so that the loop run ends when x (in cell E2) exceeds 20.

b) How do we determine y for $x = 1, 2, 3, ..., 10$?

Answer: change the formula in cell E2 so that it now reads: "=_step+1".

Do not enter 1 in cell B9, believing that the loop counter (_step) will now start with 1 instead of 0. Recall that BuildIt will always initialize the loop counter with 0 at the beginning of any model run, so whatever value we place in cell B9 will be overwritten without warning by BuildIt.

But by having "_step+1" in cell E2, x will always be one more than the loop counter. For instance, when the loop counter is 0, x is 1, and when the loop counter is incremented to 1, x is now 2, and so on.

c) How do we determine y for x=1, 3, 5, 7, 9?

Answer: like before, change the formula in cell E2 so that it now reads: "=_step+1" so that x is always one more than the loop counter. In addition, enter 2 in cell B8 so that the step size is 2 which would increment the loop counter by 2.

d) How do we determine y for x=-10, -9, ..., 9, 10?

Answer: change the formula in cell E2 so that it now reads: "=_step-10" so that x is always 10 less than the loop counter.

2.3 Example 2. Building a leaf photosynthesis model

Let us next consider a non-trivial model describing leaf photosynthesis. As given by Goudriaan and van Laar (1994), leaf photosynthesis can be described by the following equation:

$$A_L = \frac{A_m \varepsilon k I_0 \exp(-kL)}{A_m + \varepsilon k I_0 \exp(-kL)} \tag{2.3}$$

Example 2. Building a leaf photosynthesis model

where A_L is the leaf photosynthesis rate (that is, the assimilation rate of CO_2 by single leaves; μg CO_2 m^{-2} leaf s^{-1}); A_m is the maximum leaf photosynthesis rate (μg CO_2 m^{-2} leaf s^{-1}); ε is the solar radiation conversion factor (μg CO_2 J^{-1}); k is the canopy extinction coefficient for solar radiation (unitless); I_o is the solar irradiance above the canopy (W m^{-2} ground or J m^{-2} ground s^{-1}); and L is the cumulative leaf area index from the canopy top to the canopy depth being considered (m^2 leaf m^{-2} ground).

In this example, we wish to determine how A_L varies with L, with L increasing from 0, 0.5, 1, ..., 9.5 until 10 m^2 leaf m^{-2} ground.

For this task, let us take A_m as 1500 μg CO_2 m^{-2} s^{-1}; k as 0.5; I_o as 200 W m^{-2}; and ε as 12 μg CO_2 J^{-1}. All these parameters are treated as constants except for L which we will vary from 0 to 10 with an interval size of 0.5.

Like before, we will establish clear zones in the Excel worksheet for the loop section, what-to-output and output sections, and the calculation section.

Furthermore, we will also avoid hardcoding the parameter values in the formulas.

The model implementation is shown in Fig. 2.7, and besides using BuildIt cell names, we will define several other cell names (Table 2.3). Using cell names eases readability and understanding of formulas.

Fig. 2.7 reveals that the cell range B2:B5 lists the parameter values for A_m, k, I_o, and ε; cell range B8:B11 the loop section; cell range E2:E5 the calculations section; and cell range G3:H3 and G6:H6 the what-to-output and output sections, respectively.

	A	B	C	D	E	F	G	H	I
1	PARAMETERS			MODEL			TO OUTPUT		
2	A_m	1500		L	=_step		L	A_L	
3	k	0.5		n1	=Am*e*k*Io*EXP(-k*L)		=L	=AL	
4	I_0	200		n2	=Am+e*k*Io*EXP(-k*L)				
5	ε	12		A_L	=E3/E4		OUTPUT		
6									
7	CONTROL								
8	max steps	10							
9	step size	0.5							
10	step								
11	criteria	=L<=B8							
12									

Fig. 2.7. Implementation of the leaf photosynthesis model in Excel.

Table 2.3. Cell names for the leaf photosynthesis model.

Cell	Cell name	Description
B2	Am	Maximum leaf photosynthesis rate
B3	k	Canopy extinction coefficient
B4	Io	Solar irradiance above the canopy
B5	e	Solar radiation conversion factor
B9	_stepsize	BuildIt cell name for the interval or step size
B10	_step	BuildIt cell name for the loop counter
B11	_criteria	BuildIt cell name for the criteria whether to continue (TRUE) or end (FALSE) the loop run
E2	L	Leaf area index
E5	AL	Leaf photosynthesis rate
G3	_read	Marks the start of the list of parameters to include in the model output
G6	_write	Marks the start of the model output

In cell E3, the use of cell names in the formula:

33

Example 2. Building a leaf photosynthesis model
= Am*e*k*Io*EXP(-k*L)

makes the formula easier to read and more quickly understood than the following equivalent formula without use of cell names:

= B2*B5*B3*B4*EXP(-B3*E2)

where cells B2, B3, B4, B5, and E2 refer to parameters A_m, k, I_0, ε, and L, respectively.

As stated before, cell range B8:B11 is the loop section. Cell B11 holds the criteria whether to continue or end the loop run. The loop run continues as long as L (cell E2; Table 2.3) is smaller than or equal to 10 (cell B8). If L exceeds 10, the loop run ends. The step size is 0.5 (cell B9) because we wish to vary L by 0.5 from 0 to 10. Finally, cell E10 holds the loop counter, and it is maintained by BuildIt.

The calculations section is in cell range E2:E5. Cells E3 and E4 are the numerator and denominator, respectively, in Eq. 2.3. Note that the Excel function for exponential, EXP, is used in both these formulas. Following Eq. 2.3, cell E5 is the result of dividing the numerator (cell B3) with the denominator (cell B4) to give A_L. It is a good idea to breakup a long equation (*see* section 1.3) such as Eq. 2.3 into two or more parts to ease readability, as we have done here.

Lastly, the L parameter is in cell E2 which refers to the loop counter (_step). Since cell E2 refers to the loop counter, L will have starting value of 0, then 0.5, 1, 1.5, and so on until 10 (recall that _step is incremented by _stepsize, which has a value of 0.5, at the end of every loop cycle). And for each L value, cell E5 recalculates and returns the corresponding A_L value.

In the what-to-output section, which starts at cell G3 (defined with cell name _read; Table 2.3), we have instructed BuildIt that only L and A_L are to be stored, and the storage of both these model parameters would begin at cell G6 (defined with cell name _write; Table 2.3).

Once the model implementation in Excel is completed, you should click the "Start Simulation" from the BuildIt menu. Fig. 2.8 shows the model results.

	A	B	C	D	E	F	G	H
1	PARAMETERS			MODEL			TO OUTPUT	
2	A_m	1500		L	10.5		L	A_L
3	k	0.5		n1	9445.533119		10.5	6.270697598
4	I_0	200		n2	1506.297022			
5	ε	12		A_L	6.270697598		OUTPUT	
6							0	666.6666667
7	CONTROL						0.5	575.8087163
8	max steps	10					1	490.0516912
9	step size	0.5					1.5	411.3815541
10	step	10.5					2	341.0755757
11	criteria	FALSE					2.5	279.6979197
12							3	227.200046
13							3.5	183.0774586
14		1000					4	146.537035
15		800					4.5	116.6437415
16		600					5	92.43216316
17							5.5	72.98101718
18		400					6	57.45602828
19		200					6.5	45.12917111
20							7	35.38210257
21		0					7.5	27.70013916
22							8	21.66137316
23							8.5	16.92395489
24							9	13.21336606
25							9.5	10.31067042
26							10	8.04218614

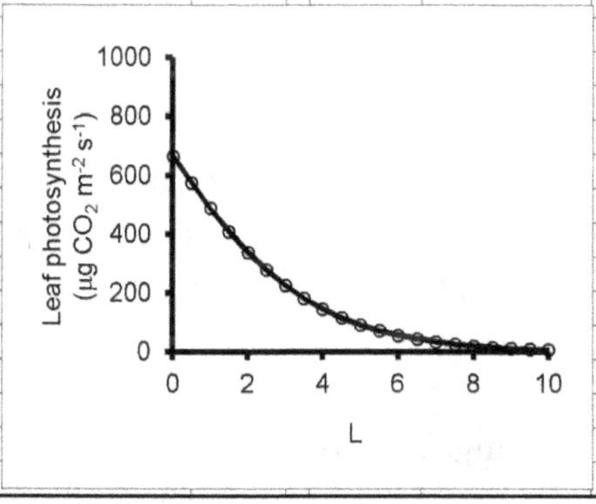

(Chart in cells A12:F26 — Leaf photosynthesis ($\mu g\ CO_2\ m^{-2}\ s^{-1}$) versus L)

Fig. 2.8. Simulation results of the leaf photosynthesis in Excel.

Cell range G6:H26 hold the model output, and this output are plotted in a chart to show how A_L varies with L. Notice that the model output is a list of (L, A_L) pairs, with L increasing from 0 to 10 at every 0.5 interval. Recall

Example 2. Building a leaf photosynthesis model

that the step size (cell B9) was set at 0.5, so that A_L could be determined for $L = 0, 0.5, 1, ..., 9.5, 10$.

At the end of the model run, the loop counter (cell B10) is 10.5, a value larger than 10, so the loop condition or criteria (cell B11) returns FALSE, a value BuildIt responds by ending the loop run.

2.3.1 Using a non-constant canopy extinction coefficient

The parameters A_m, k, I_o, and ε in Eq. 2.3 were taken as constants in the previous example. However, we will now consider a non-constant canopy extinction coefficient, k.

From Teh (2006), the canopy extinction coefficient for the *diffuse* component of solar radiation can be computed as:

$$k = \frac{1+0.1174\sqrt{L}}{1+0.3732\sqrt{L}} \qquad (2.4)$$

where L, as before, is the cumulative leaf area index.

Note, for the sake of simplicity, we will consider the canopy extinction coefficient only for the diffuse component of solar radiation. But in practice, the direct component of solar radiation must also be considered.

Fig. 2.9 shows the changes in the implementation to accommodate the use of a non-constant k in the model.

Previously, k, being a constant, was stored in cell B3 (Fig. 2.7), but in keeping with our practice to demarcate areas in the worksheet for loop, calculations, and output, k is moved to cell E5 in the calculations section(Fig. 2.9).

Cell E6 (previously E5 in Fig. 2.7) now contains the calculated value for A_L.

	A	B	C	D	E	F
1	PARAMETERS			MODEL		
2	A_m	1500		L	=_step	
3				n1	=Am*e*k*Io*EXP(-k*L)	
4	I_0	200		n2	=Am+e*k*Io*EXP(-k*L)	
5	ε	12		k	=(1+0.1174*SQRT(L))/(1+0.3732*SQRT(L))	
6				A_L	=E3/E4	
7	CONTROL					
8	max steps	10				
9	step size	0.5				
10	step					
11	criteria	=L<=B8				
12						

Fig. 2.9. Modifying the leaf photosynthesis model to include the computation for k (canopy extinction coefficient).

Please note that since the cells for computing k and A_L have been moved, the definition of cell names k and AL should likewise move to their new cell locations (Table 2.4).

Table 2.4. New cell locations for the cell names k and AL after the leaf photosynthesis model is modified to have a non-constant k (canopy extinction coefficient).

Cell name	Previous cell	New cell
k	B3	E5
AL	E5	E6

The what-to-output section is also slightly modified because we want to additionally see the k values in the model output (Fig. 2.10 and 2.11). The what-to-output section now contains three items to be included in the model output: L, k, then A_L.

With all these changes, the model output after a model run should look like Fig. 2.11.

Example 2. Building a leaf photosynthesis model

	F	G	H	I	J
1		TO OUTPUT			
2		L	k	A_L	
3		=L	=k	=AL	
4					
5		OUTPUT			
6					

Fig. 2.10. The what-to-output section is modified to include the output of k (canopy extinction coefficient).

	A	B	C	D	E	F	G	H	I	J
1	PARAMETERS			MODEL			TO OUTPUT			
2	A_m	1500		L	10.5		L	k	A_L	
3				n1	3182.9		10.5	0.62	2.12	
4	I_0	200		n2	1502.1					
5	ε	12		k	0.6248		OUTPUT			
6				A_L	2.1189		0	1	923	
7	CONTROL						0.5	0.86	708	
8	max steps	10					1	0.81	549	
9	step size	0.5					1.5	0.78	418	
10	step	10.5					2	0.76	315	
11	criteria	FALSE					2.5	0.75	234	
12							3	0.73	173	
13							3.5	0.72	128	
14							4	0.71	94	
15							4.5	0.7	69.3	
16							5	0.69	51.1	
17							5.5	0.68	37.8	
18							6	0.67	28	
19							6.5	0.67	20.8	
20							7	0.66	15.5	
21							7.5	0.65	11.6	
22							8	0.65	8.67	
23							8.5	0.64	6.51	
24							9	0.64	4.9	
25							9.5	0.63	3.7	
26							10	0.63	2.8	
27										

Fig. 2.11. Model output of the modified leaf photosynthesis model.

With increasing L, notice that k is not constant but declines from 1.0 to 0.63 and that A_L declines at a faster rate than that when k was fixed at 0.5 (compare Fig 2.8 with Fig. 2.11).

In the next chapter, we will delve deeper into BuildIt's features.

2.4 Exercises

1. Using BuildIt, implement the following equations in Excel:

   a) $y = 2\sin(x^2)$ where x is in unit radian.

      Determine y for x = 0 to π at every $\pi/10$ radian interval.

   b) $y = x^{0.9} + 2^{-x}x$

      Determine y for x = 0 to 1 at every 0.1 unit interval.

   c) $y = \exp(-x)$

      Determine y for x = -2 to 2 at every 0.5 unit interval.

   d) $y = \dfrac{2x + 3}{x^2 + 2}$

      Determine y for x = -10 to 10 at every 1 unit interval.

   For each equation, the output should be a list of (x, y) pairs, and use Excel to plot these (x, y) pairs in a scatter chart.

2. Following the scenarios presented in section 2.2.2.5, what changes in the model implementation are required if we wish to determine y for

a) x = 2, 4, 6, 8, 10?

b) x = 0, 2, 6, 14, 30, 62?

c) x = -2, -4, -6, -8, -10 (in this order)?

d) x = 5, 4, 3, 2, 1, 0 (in this order)?

3. Ehleringer and Björkman (1977) determined that the solar radiation conversion factor ε (μg CO_2 J^{-1}) for C3 grasses varied with canopy temperature T_f (°C) according to:

$$\varepsilon = 16.2162 - 0.010620 T_f - 0.003804 T_f^2$$

Use this equation in Eq. 2.3 to determine A_L, the leaf photosynthesis rate, for leaf area index L = 0 to 10 m^2 leaf m^{-2} ground at every 0.5 m^2 leaf m^{-2} ground interval. Take A_m as 1500 μg CO_2 m^{-2} s^{-1}, k as 0.5, I_0 as 200 W m^{-2}, and T_f as 25 °C.
Determine your answers using BuildIt.

4. From Question 3, use BuildIt to determine A_L for T_f = 10 to 40 °C at every 5 °C interval. Unlike before, the leaf area index L is fixed at 2 m^2 leaf m^{-2} ground. For the other parameters, use their previous values: A_m as 1500 μg CO_2 m^{-2} s^{-1}, k as 0.5, and I_0 as 200 W m^{-2}.
Plot the model output: A_L against T_f. How does A_L vary with T_f?

Chapter 3. BuildIt actions (Part 1)

3.1 Actions

Whenever we implement a model in Excel, we often require certain spreadsheet cell operations to occur. But some of these cell operations can either be too difficult to do in native Excel or require us to directly use Excel VBA (Excel's programming language).

One cell operation often needed, for instance, is for one cell to change or update the value of another cell. But in Excel, this task is not allowed. Cell A1, for example, can only read but not change the content in cell B1. From cell A1's perspective, cell B1 as well as other external cells are strictly read-only.

Another notable cell operation often required is for a cell to update its own current value. For instance, consider the following two innocuous equations:

$$x_{i+1} = x_i + a_i$$

and

$$x_{i+1} = x_i \times a_i$$

where x is updated by adding or multiplying its current value with the value a.

In Excel, x_i and x_{i+1} must occupy separate cells. This is because a cell cannot update itself by referring to its current values. There are no Excel formulas or operations where cell A1, for instance, can take its own current value, add (or multiply) it with 10, then update itself with this new value. In other words, having the following formulas in cell A1: "=A1+10" or "=A1*10" would be invalid because of so-called circular reference where cell A1 refers to itself. Besides, having self-referential cells would cause never-ending cell updates.

To overcome these two and other limitations of Excel, BuildIt supplies 12 *actions*. These so-called actions perform specific tasks that would otherwise be impossible to carry out in native Excel. Table 3.5 lists the 12 BuildIt actions, and how they can be used will be discussed shortly.

Table 3.5. BuildIt actions

Code	Action name	Brief description
ITG	Integration	Numerical integration using the five-point Gaussian method.
DIF	Differentiation	Numerical differentiation to determine either the first or second derivative using the three-point central (midpoint) difference method.
ROT	Rotate	Rotates the order in an array of values to the left, right, up, or down direction. For example, given an array of (1, 2, 3, 4, 5), rotating it to the right or down would give (5, 1, 2, 3, 4), but rotating it to the left or up would give (2, 3, 4, 5, 1).
REV	Reverse	Reverses the order in an array of values. For example, given an array of (1, 2, 3, 4, 5), reversing it would give (5, 4, 3, 2, 1).

Code	Action name	Brief description
SHU	Shuffle	Randomly shuffles the order in an array of values. For example, given an array of (1, 2, 3, 4, 5), a shuffle might give (3, 4, 1, 2, 5); the sequence randomly determined.
SOR	Sort	Sorts the order in an array of values either ascendingly or descending.
ARR	Arrange	Arranges the order in an array of values according to their given sequence. For example, given an array of (10, 20, 30, 40, 50) and the corresponding sequence as (3, 1, 4, 5, 2), this action would give the new array of values as (20, 50, 10, 30, 40).
ACC	Cumulative	Determines the cumulative sum or product of a given variable.
REP	Copy value	Copies the values in source cells to destination cells.
UPD	Update value	Updates the values of one or more variables by their rates of change using the Euler method.
INI	Initialize value	Initializes one or more variables with their given initial values.

Code	Action name	Brief description
RUN	Runs a macro or script	Runs a user-defined VBA script or macro. BuildIt supplies three utility macros: 1) `ClearOutput`, 2) `EnableScreenUpdate`, and 3) `DisableScreenUpdate`.

The specification of a BuildIt action always starts with a three-letter code (Table 3.5), followed by two or more parameters, depending on the type of action (Fig. 3.1). And to instruct BuildIt where these actions reside in the worksheet, we need to define a cell with the cell name `_operation`. This cell marks the start of the list of actions BuildIt will read and execute.

In Fig. 3.1, cell name G2 is defined with the name `_operation`, so BuildIt will execute the actions listed there, starting at cell G2. There are four actions listed: one ITG, followed by two UPD, and the last action: ACC. BuildIt reads and execute these four actions according the order they are listed: starting with the ITG action and ending with the ACC action.

	G	H	I	J	K	L	M	N	O
1	OPERATION								
2	ITG	=A1	=A2	=A3	1	5	TRUE	TRUE	
3	UPD	=D7	=E7	1	TRUE				
4	UPD	=D8	=E8	4	TRUE				
5	ACC	=D12	=D13	*	TRUE				
6									
7									

Fig. 3.1. A list of four BuildIt actions, starting from cell G2 which has been defined with the cell name `_operation`.

Each action must be specified starting with its three-letter code (ITG, UPD, or ACC), followed by its parameters. These parameters contain information pertaining to the action's task. ITG action has seven parameters (cell range H2:N2), UPD four parameters (H3:K3 for the first UPD and H4:K4 for the second), and ACC four parameters (H5:K5).

Each BuildIt action is specified in a single cell row, so that the first blank cell row marks the end of the action list. In Fig. 3.1, notice that cell G6 and the cells to its right are blank, so the action list ends with the ACC action (G5:K5).

All BuildIt actions have the following syntax:

```
ACTION    {param₁:s|r}
          {param₂:s|r}
          ...
          {paramₙ:s|r}
          {op:s=TRUE}
```

where ACTION denotes a three-letter code for the type of action requested (Table 3.5), and $param_{1,2,...,n}$ are the action's first, second, ..., and the *n*-th parameters, where the number of parameters depends on the type of action.

Each parameter's ":s|r" notation means a parameter can either be of type ":s" or ":r". The ":s" type means the parameter must be supplied either as a constant (such as numbers, logical values TRUE and FALSE, or text) or a single cell address (such as A1 and $B$5). The ":r" type is the same as ":s" except it can additionally take a contiguous cell range (such as A1:B3 and C4:C6).

Note that the last parameter op is unique because this parameter is always of type ":s" and it only accepts logical values TRUE or FALSE. The last parameter indicates whether or not the given action should be executed. If op is TRUE, the action is executed, but a FALSE value means the action will be not be carried out. This last parameter gives the flexibility of executing a given action only in certain conditions. If the last parameter is unspecified, it is TRUE by default, which explains the notation "op:s=TRUE".

As usual, we will consider several examples to better understand how to use BuildIt actions.

3.2 Integration

Use Excel to solve:

$$A = \int_{1}^{5} \left(2x^3 + 4x + 1\right) dx \qquad (3.1)$$

where the function $2x^3 + 4x + 1$ is integrated with respect to x between the lower and upper limits of 1 and 5, respectively.

To perform integrations, the BuildIt action ITG should be used (Table 3.5) which uses the five-point Gaussian numerical integration method. In most cases, this integration method is suffice and will give accurate solutions.

Fig. 3.2 shows how the ITG action is specified in the worksheet. The function $2x^3 + 4x + 1$ is implemented in cell B2 with the equivalent Excel formula: "$= 2{*}B1^3 + 4{*}B1 + 1$" where cell B1 holds the x value.

	A	B	C	D	E	F	G	H	I	J	K	L
1	x			OPERATION								
2	f(x)	=2*B1^3+4*B1+1		*Integrate*	**ITG**	=B1	=B2	=B3	1	5	TRUE	
3	A											
4												

Fig. 3.2. Using the ITG action to perform integrations in Excel.

Cell E2 is defined with the cell name `_operation`. This cell marks the start of actions BuildIt will read and execute. However, in this example, only one action is specified. Like in cells A1, A2, and A3, the text in cell D2 is merely a label. Such labels are optional although recommended to aid model readability and for documentation purpose.

The `ITG` action has the following syntax:

```
ITG        {x:s}
           {fn:s}
           {output:s}
           {lower:s}
           {upper:s}
           {single_integral:s=TRUE}
           {op:s=TRUE}
```

This action integrates the function `fn` of `x` over the interval [`lower`, `upper`], and the result of the integration is to be stored in `output` cell. The second last parameter, `single_integral`, is either TRUE for single integrations or FALSE for multiple integrations (more about multiple integrations later). If omitted, `single_integral` is assumed TRUE by default. The last parameter `op` should be set to TRUE for BuildIt to execute this action; otherwise set FALSE to cancel its execution. `op` is assumed TRUE if it is unspecified (left blank).

A BuildIt action and its parameters must be specified in a single row. In Fig. 3.2, `ITG` and its seven parameters are specified in a single row from cell E2 to L2 (recall cell D2 is only a text label and not part of the action's syntax). Cell E2 is the three-letter code `ITG` so that BuildIt understands the action for integration is chosen. The subsequent cells F2 to L2 are the `ITG`'s seven parameters: cell F2 is for `x`, G2 for `fn`, H2 for `output`, I2 for `lower`, J2 for `upper`, K2 for `single_integral`; and the blank cell L2 for `op`.

The function to be integrated is specified in cell H2 which refers to cell B2. This function is to be integrated over the intervals 1 and 5 (specified in cells I2 and J2, respectively), and the location for storing the result of this integration is specified in cell H2 which refers to cell B3. In other words, BuildIt will store the integration result (A in Eq. 3.1) in cell B3. Cell F1 refers to cell B1 which holds the value for x. This cell should be left empty or unspecified because BuildIt will use cell B1 in its internal calculations to solve Eq. 3.1. If cell B1 has a value, it will be overwritten without warning by BuildIt. Cell K2 is specified TRUE because Eq. 3.1 is a single integral. This cell can also be left blank which BuildIt will take as TRUE by default. Lastly, cell L2 is left blank, so the ITG's last parameter op is assumed TRUE. So long as op is TRUE BuildIt will execute this ITG action.

Note that there is no need to define cell names _step, _stepsize, and _criteria as taught in the previous chapter. This is because solving Eq. 3.1 does not require a loop for repetitious calculations. There is also no need to define cell names _read and _write because there is no output listing. A set of calculations is performed only once in this example, and the output is a single result stored in a single cell: cell B3.

When implementation of Eq. 3.1 in Excel is complete, you should click the "Do Operations Only" from BuildIt menu. Do not click "Start Simulation" because this command is only for models requiring iterative calculations. Choosing "Start Simulation" in this example will result in BuildIt displaying an error message that the cell names (step, _stepsize, and _criteria) for running the loop are missing.

Fig. 3.3 shows the result after the "Do Operations Only" command is selected.

	A	B	C	D	E	F	G	H	I	J	K	L
1	x	4.81236		OPERATION								
2	f(x)	243.146		*Integrate*	ITG	4.812	243.1	364	1	5	TRUE	
3	A	364										
4												

Fig. 3.3. The integration result.

Cell B3 holds the value 364 which is the integration result for Eq. 3.1. Notice that upon completion of calculations, cell B1 holds the value 4.81 which was used as part of BuildIt's internal calculations to solve Eq. 3.1. This value (as well as the value in cell B2) can be ignored.

What about double integrations? Consider the following problem:

$$A = \int_0^2 \int_1^5 \left(2x^3 + 4xy + 1\right)dxdy \tag{3.2}$$

To perform double integrations, two ITG actions need to be specified in succession. Fig. 3.4 shows how this can be implemented in a worksheet.

	A	B	C	D	E	F	G	H	I	J	K	L
1	x			OPERATION								
2	y			Integrate x	ITG	=B1	=B3	=B4	1	5	FALSE	
3	f(x)	=2*B1^3+4*B1*B2+1		Integrate y	ITG	=B2	=B3	=B4	0	2	TRUE	
4	A											

Fig. 3.4. Two successive ITG actions to perform a double integration in Excel.

Cell B3 implements the function $2x^3 + 4xy + 1$ using x and y values from cells B1 and B2, respectively.

As before, cell name _operation is defined for cell E2. This instructs BuildIt to start reading and executing the list of actions starting in cell E2. Two actions are specified: the first ITG action in cell range E2:L2 and the second ITG action in E3:L3. Recall that each BuildIt action must be specified in a single row.

Two ITG actions must be specified, and they must be specified in succession (*i.e.*, the second ITG right after the first ITG) because solving Eq. 3.2 requires two successive integrations: the first with respect to x (cell range E2:L2), then with respect to y (cell range E3:L3).

Notice that cells K2 and K3 are set to FALSE and TRUE, respectively. Both these cells hold the value for the single_integral parameter for the ITG action. Cell K2 is set to FALSE to instruct BuildIt that the next integration should also be performed on the same given function (in cell B3). For the second integration, cell K3 is set to TRUE (or can be left unspecified) because this second ITG action would be the final integration to be performed on the given function.

Also notice that cell F2 (parameter x) refers to cell B1 which holds the x value. This is done because the first ITG action is to integrate the given function with respect to x, but the second integration is with respect to y, so cell F3 refers to cell B2 which holds the y value.

As before, you should click the "Do Operations Only" from the BuildIt menu to solve Eq. 3.2.

The result of the double integration is 728, and it is stored in cell B4 (Fig. 3.5).

	A	B	C	D	E	F	G	H	I	J	K	L
1	x	4.81236		OPERATION								
2	y	1.90618		Integrate x	ITG	4.81	261	728	1	5	FALSE	
3	f(x)	260.5899		Integrate y	ITG	1.91	261	728	0	2	TRUE	
4	A	728										
5												

Fig. 3.5. The double integration result in cell B4.

3.3 Differentiation

Differentiation in Excel can be performed using the DIF action (Table 3.5). Consider the following example:

$$y = 2x^3 + 1 \qquad (3.3)$$

We wish to determine the first derivative of y with respect to x; that is, to determine dy/dx.

Fig. 3.6 shows how the DIF action is specified in a worksheet. Cell B2 implements Eq. 3.3 using the x value in cell A1. In this case, x is set to 2. Cell name _operation is defined for cell E2 so that BuildIt knows where the list of actions is located in the worksheet. In this case, only one action is specified in cell range E2:K2.

	A	B	C	D	E	F	G	H	I	J	K
1	x	2		**OPERATION**							
2	y	=2*B1^3+1		Differentiate	**DIF**	=B1	=B2	=B3			
3	dy/dx										
4											

Fig. 3.6. Using the DIF action for numerical differentiation.

The DIF action has the following syntax:

```
DIF        {x:s}
           {fn:s}
           {output:s}
           {n:s=1}
           {interval_size:s=0.001}
           {op:s=TRUE}
```

which gives the first or second derivative of $d^n fn/dx^n$, where n is either 1 (first derivative) or 2 (second derivative). By default, n is 1. The interval_size for numerical differentiation is 0.001 by default, but this value may require some experimentation as too small or large a value may give large errors. The output parameter is the location where the result of differentiation should be stored, and the op parameter, as before, denotes if this DIF action should be executed (TRUE if to execute the action, else FALSE to cancel the execution).

In Fig. 3.6, cell E2 holds the three-letter code to instruct BuildIt to execute the DIF action. The subsequent cells F2 to H2 instruct BuildIt to differentiate the given function (cell B2) with respect to *x* (cell B1), and the result is to be stored in cell B3.

Cells I3, J3, and K3 are left blank, so their respective default values are assumed. Since only the first derivative *dy/dx* is required, cell I3 (parameter n) should be 1 (or left blank). To determine the second derivative, enter 2 in cell I3. Cell K3 is also left blank (so, op=TRUE by default), so this DIF action will always be performed.

Once implementation is complete, click the "Do Operations Only" from the BuildIt menu.

With *x*=2, the first derivative of Eq. 3.3 is 24, as seen in Fig. 3.7.

	A	B	C	D	E	F	G	H	I	J	K
1	x	2		**OPERATION**							
2	y	17		Differentiate	**DIF**	2	17	24			
3	dy/dx	24									
4											

Fig. 3.7. The first derivative result (cell C3) when *x* is 2.

It is important to remember that, unlike cell B2, cell B3 does not hold a formula, so if the value of *x* (cell B1) changes, you must click again the "Do Operations Only" from the BuildIt menu to recalculate and refresh the value in cell B3.

3.4 Actions for array (list) manipulations

BuildIt supplies five actions to manipulate an array (list) of numbers: ROT, REV, SHU, SOR, and ARR (Table 3.5).

The ROT action, for instance, takes an array of numbers and moves each number in the array in a specified direction: to the left, right, up, or down.

Fig. 3.8 shows a list of five numbers (1 to 5) arranged vertically from cell A2 to A6. Cell D2 is defined with the cell name _operation, so that BuildIt knows where to find the list of actions (only one action is listed in this case).

	A	B	C	D	E	F	G
1	**LIST**		**OPERATION**				
2	1		Rotate	**ROT**	=A2:A6	DOWN	
3	2						
4	3						
5	4						
6	5						
7							

Fig. 3.8. ROT action moves a given array of numbers downward.

ROT action has the following syntax:

```
ROT        {array_of_values:r}
           {direction:s=RIGHT}
           {op:s=TRUE}
```

which rotates the order in array_of_values (*i.e.*, list of values) in the direction specified by LEFT, RIGHT, UP, or DOWN.

For example, given an array of (1, 2, 3, 4, 5), rotating it RIGHT or DOWN would give (5, 1, 2, 3, 4). Instead, rotating it LEFT or UP would give (2, 3, 4, 5, 1).

Directions LEFT and RIGHT are given for arrays specified in a single cell row, such as cell range A1:F1. Directions UP and DOWN are for arrays specified in a single cell column, such as in cell range A1:A6.

Click the "Do Operations Only", and Fig. 3.9a shows the result of the array rotation.

Notice that after rotation, each number in the array has been moved one cell downward. Number 1 has moved from cell A2 to A3, and likewise for number 2 from cell A3 to A4. Number 5 was wrapped round the array, so that it now resides in cell A2, having moved from cell A6.

a)

	A	B	C	D	E	F
1	LIST		OPERATION			
2	5		Rotate	ROT	5	DOWN
3	1					
4	2					
5	3					
6	4					

b)

	A	B	C	D	E	F
1	LIST		OPERATION			
2	4		Rotate	ROT	4	DOWN
3	5					
4	1					
5	2					
6	3					

Fig. 3.9. An array of numbers (cells A2:A6) after: a) one and b) two downward rotations.

Click the "Do Operations Only" once more, and the numbers in the array are moved one cell downward again

(Fig. 3.9b). Specifying the `direction` parameter (in cell F2) UP will move the numbers in the opposite direction. An array (1, 2, 3, 4, 5) rotated UP will give (2, 3, 4, 5, 1), where 1 is wrapped round to occupy the last position in the array.

Arrays can also be arranged horizontally, such as shown in Fig. 3.10.

a)

	A	B	C	D	E
1	LIST				
2	1	2	3	4	5
3					
4	OPERATION				
5	Rotate	**ROT**	=A2:E2	RIGHT	
6					
7					

b)

	A	B	C	D	E
1	LIST				
2	5	1	2	3	4
3					
4	OPERATION				
5	Rotate	**ROT**	=A2:E2	RIGHT	
6					
7					

Fig. 3.10. ROT action: a) horizontally arranged array (1, 2, 3, 4, 5) is to be shifted one cell to the right, and b) the result after one rotation. Note: cell B5 is defined with the cell name _operation.

From Fig. 3.10, clicking the "Do Operations Only" will rotate the numbers in the given array (1, 2, 3, 4, 5) once to the right, giving (5, 1, 2, 3, 4) in cell range A2:E2.

Again, the last number 5 is wrapped round from cell E2, so that this number now resides in the first cell in A2.

Specifying the `direction` parameter (in cell D5) LEFT will move the numbers to the left, so that the result is instead (2, 3, 4, 5, 1), where the first number 1 is wrapped round from cell A2 to E2.

The other four array actions, REV, SHU, SOR, and ARR, are used and they work similarly to the ROT action. Given an array of numbers, these actions will move each of these numbers in a particular way. The syntax for these four array-manipulation actions are as follows:

a)   REV action

```
REV    {array_of_values:r}
       {op:s=TRUE}
```

reverses the order in `array_of_values`. For example, given an array of (1, 2, 3, 4, 5), reversing it would give (5, 4, 3, 2, 1). Reversing it again produces the original sequence.

b)   SHU action

```
SHU    {array_of_values:r}
       {op:s=TRUE}
```

shuffles the order in `array_of_values` at random. For example, given an array of (1, 2, 3, 4, 5), a shuffle might give (3, 4, 1, 2, 5), where the sequence is randomly determined by Excel's random number generator.

c)   SOR action

```
SOR    {array_of_values:r}
       {ascending_order:s=TRUE}
       {op:s=TRUE}
```

sorts the `array_of_values` either in an ascending or descending order. Specify `ascending_order` as TRUE to sort ascendingly the values in a list such as (3, 1, 5, 2, 4) to give (1, 2, 3, 4, 5), or else set to FALSE for a descending order to give (5, 4, 3, 2, 1).

56

d) ARR action

```
ARR    {array_of_values:r}
       {array_of_sequence:r}
       {op:s=TRUE}
```

arranges `array_of_values` in the order specified by `array_of_sequence`. For example, take an array of (10, 20, 30, 40, 50) and the `array_of_sequence` given as (3, 1, 4, 5, 2) (Fig. 3.11).

a)

	A	B	C	D	E	F	G	H
1	LIST	ORDER		OPERATION				
2	10	3		Arrange	ARR	=A2:A6	=B2:B6	
3	20	1						
4	30	5						
5	40	4						
6	50	2						
7								

b)

	A	B	C	D	E	F	G	H
1	LIST	ORDER		OPERATION				
2	20	3		Arrange	ARR	=A2:A6	=B2:B6	
3	50	1						
4	10	5						
5	40	4						
6	30	2						
7								

Fig. 3.11. ARR action: a) an array of numbers (in cells A2:A6) is to be arranged according to the order specified in cells B2:B6, and b) the results after arrangement. Note: cell E2 is defined with the cell name _operation.

Arranging the array would give the new array as (20, 50, 10, 30, 40), where 10 corresponds to sequence 3, 20 to 1, 30 to 4, 40 to 5, and 50 to 2.

Note that the cell range specified for `array_of_values` and `array_of_sequence` do not need to be in the same horizontal or vertical arrangement. For instance, cell range A2:A6 can be the `array_of_values` (a vertical cell range) and D6:H6 the `array_of_sequence` (a horizontal cell range).

3.5   Using BuildIt actions in iterative calculations

BuildIt actions become much more useful when they are used in a loop for iterative calculations. Fig. 3.12 expands the program flow of BuildIt which includes the execution of actions in the operation section within the loop.

In the previous chapter, we learn that at the start of a simulation run, BuildIt will always initialize the loop counter `_step` with 0, after which the loop criteria `_criteria` is checked. If `_criteria` is TRUE, the loop begins.

After all model calculations are completed, BuildIt finds the cell defined with the cell name `_operation` because this cells marks the start of the list of actions BuildIt will execute. Once all of the actions are executed, the selected model results are stored. The loop counter `_step` is incremented by the interval size `_stepsize`. The loop condition `_criteria` is checked again. The loop and model run ends if `_criteria` is FALSE.

Please note that executing one or more actions will often trigger a fresh round of calculations in the model as one or more cells will be updated with new values. Only when all calculations have been completed will BuildIt output the selected model results.

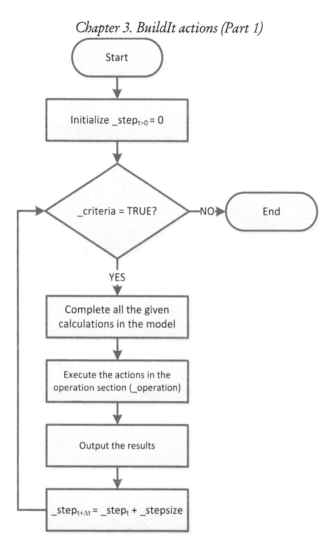

Fig. 3.12. Actions are read and executed by BuildIt after all calculations in the model are completed in the loop. However, executing the actions may trigger a fresh round of calculations by Excel.

This section will now demonstrate how BuildIt actions can be used within a loop.

3.5.1 Cumulative sum and product

Some BuildIt actions are meaningful only when they are used within a loop. The ACC action is one of them. This action is used to determine the cumulative sum or product (Table 3.5).

Consider the following general equation to determine the factorial of a number n :

$$n! = n \times (n-1) \times (n-2) \times ... \times 1 \qquad (3.4)$$

For example, the factorial of 5 (or 5!) is:

$$5! = 5 \times 4 \times 3 \times 2 \times 1$$
$$= 120$$

and 10! is

$$10! = 10 \times 9 \times 8 \times 7 \times 6 \times 5 \times 4 \times 3 \times 2 \times 1$$
$$= 3,628,800$$

Although using native Excel to determine factorial numbers is very easy, we will use Excel in conjunction with BuildIt just to demonstrate one scenario in which the ACC action can be applied.

Fig. 3.13 shows one way how to implement Eq. 3.4 in Excel.

Cell B2 holds the factorial number we wish to calculate, which, in this case, is 5. This cell is also defined with the cell name nfactorial (Table 3.6).

Cells B5, B6, and B7 hold information about the loop: cell B6 is the loop counter (defined with cell name _step); B5 the step size (_stepsize); and B7 the criteria on whether to continue or terminate the loop run (_criteria) (Table 3.6).

	A	B	C	D	E	F	G	H	I	J	K
1	INPUT			MODEL			OPERATION				
2	factorial	5		n	=nfactorial-_step		ACC	=n	=E3	*	TRUE
3				n!							
4	CONTROL										
5	stepsize	1									
6	step										
7	criteria	=n>=1									
8											

Fig. 3.13. Using ACC action to determine factorial numbers.

Table 3.6. Cell names defined in the worksheet to determine factorial numbers (BuildIt cell names are listed first).

Cell	Cell name
B5	_stepsize
B6	_step
B7	_criteria
H2	_operation
B2	nfactorial
E2	n

Cell H2 marks the start of the list of actions BuildIt will perform, so this cell is defined with the cell name _operation. In this example, only one action is requested: the ACC action, which has the following syntax:

```
ACC     {x:s}
        {cumulative_x:s}
        {operator_type:s}
        {op:s=TRUE}
```

which obtains the cumulative sum of x if operator_type is + or the cumulative product of x if operator_type is *. The cumulative sum or product is stored in the cell specified in cumulative_x. As always, the last parameter op is TRUE by default which means this action will be executed by BuildIt, else set to FALSE to cancel the execution.

The cells G2:K2 define the ACC action, where cell G2 is the three-letter code for the ACC action and cell H2 refers to n, the cell name for E2 (Table 3.6). Cell I3 means the cumulative result is to be stored in cell E3, and cell J2 means the cumulative product of *n* is required (see Eq. 3.4). Lastly, cell K3 is set to TRUE (or can be left blank) to mean the ACC action is always to be executed.

These ACC parameters instruct BuildIt to multiply n with the value stored in cell E3. Their product is stored in cell E3. In other words, cell E3 is updated by multiplying its current value with n (from cell E2). This update occurs once every loop cycle (Table 3.12).

The value of n is not constant because its value changes in every loop cycle. Notice in Fig. 3.13 that cell E2 (n) contains the formula "=nfactorial-_step". At the start of a model run, BuildIt will initialize the loop counter _step with 0 (Fig. 3.12), so when the loop begins, n is (5-0) = 5, then 4, 3, 2, and 1 in successive loop cycles. When n reaches (5-5) = 0, the loop run ends. The loop ends at n = 0 because of the loop criteria in cell B7 (_criteria) which contains the formula "=n> =1".

In other words, the ACC action allows cell E3 to store the result of $(5 \times 4 \times 3 \times 2 \times 1)$ for the case of 5! (Fig. 3.14a) and $(10 \times 9 \times 8 \times 7 \times 6 \times 5 \times 4 \times 3 \times 2 \times 1)$ for 10! (Fig. 3.14b).

Once model implementation is completed, click the "Start Simulation" command from the BuildIt menu. Choosing this command will run the required iterative calculations. Clicking the "Do Operations Only" from the BuildIt menu will only cause the ACC action to be executed and its execution is done only once. This will not give us the correct result for the factorial calculations.

a)

	A	B	C	D	E	F	G	H	I	J	K
1	INPUT			MODEL			OPERATION				
2	factorial	5		n	0		ACC	0	120	*	TRUE
3				n!	120						
4	CONTROL										
5	stepsize	1									
6	step	5									
7	criteria	FALSE									
8											

b)

	A	B	C	D	E	F	G	H	I	J	K
1	INPUT			MODEL			OPERATION				
2	factorial	10		n	0		ACC	0	4E+06	*	TRUE
3				n!	3628800						
4	CONTROL										
5	stepsize	1									
6	step	10									
7	criteria	FALSE									
8											

Fig. 3.14. Result of the ACC action: Factorial of a) 5 and b) 10.

Note that if cell J2 (parameter operator_type) is replaced by +, the cumulative sum (*e.g.*, 5 + 4 + 3 + 2 + 1 or 10 + 9 + 8 +7 + 6 + 5 + 4 + 3 + 2 + 1) is obtained instead.

The ACC action is one way a cell can change the contents of another cell. The value of cell E3 (Fig. 3.13) is updated by multiplying (or adding) its current value by another value from cell E2. This kind of operation is impossible to do in native Excel – or possible only by using Excel's programming language.

Besides ACC, three other BuildIt actions, INI, UPD and REP, also allow us to update or change the contents of other cells.

3.5.2   Cell copy operation

The ACC action, as previously discussed, deals with the specific task of calculating either the cumulative sum or product of a variable. The REP action (Table 3.5) is more generic: it copies the contents of one cell to a target cell. The REP action is often used to update the value of a specific cell.

Let us consider the following example:

$$a_i = \begin{cases} 3 & i = 1 \\ 2a_{i-1} + 5 & i > 1 \end{cases} \qquad (3.5)$$

so that $a_1 = 3$, $a_2 = 2a_1 + 5$, $a_3 = 2a_2 + 5$, $a_4 = 2a_3 + 5$, and so on. We can see that, other than for a_1, determining the value for a_n requires that we recursively determine the values for a_{n-1}, a_{n-2}, a_{n-3}, and so on until a_1.

In this example, we wish to determine the value for a_5. Implementing Eq. 3.5 in native Excel is straightforward, but it requires the use of more than one spreadsheet cell to hold the various a_i values; that is, a_1, a_2, a_3, ..., a_n must exist separately in their respective cells. Using the REP action however allows us to use a single cell to store the latest value of a_i and allows updates to occur within this same cell.

Fig. 3.15 shows the use of REP to calculate a_5 based on Eq. 3.5.

a)

	A	B	C	D	E	F
1	INPUT			MODEL		
2	n	5		i	=_step+1	
3				a_{i-1} (or a_n)	123	
4	CONTROL			a_i	=IF(E2=1,3,2*E3+5)	
5	stepsize	1				
6	step					
7	criteria	=E2<=B2				
8						

b)

	F	G	H	I	J	K
1		OPERATION				
2		REP	=E3	=E4		
3						
4						

Fig. 3.15. Use of the REP action in Excel to recursively determine the value of a_5, where $a_i = 2a_{i-1}+5$ and $a_1 = 3$. Setup of the a) loop and calculation sections, and b) the operation section.

Cell B2 denotes $n = 5$ because we wish to determine the value of a_5. Cells B5:B7 contain the information on the loop for BuildIt. Cell B5, B6, and B7 are defined with the cell names _stepsize, _step, and _criteria, respectively. The loop run continues as long as i (cell E2) is smaller than or equal to n (cell B2). Cell E2 contains the formula "=_step+1" so that i successively increases at every loop cycle from 1 to 2, then to 3, 4, and so on until the loop ends when $i > n$ (cell B7). Recall that _step is always initialized with 0 at the start of a simulation or model run.

Cell E3 and E4 hold the a_{i-1} and a_i values, respectively. Cell E4 in particular contains the formula "=IF(E2=1,3,2*E3+5)". The logical IF function is used to implement Eq. 3.5, so that when $i = 1$, $a_1 = 3$, else $a_i = 2a_{i-1}+5$ for $i > 1$, where the value for a_{i-1} is taken from cell E3.

Cell E4 uses information from cell E3 to update the a_i value. The next step is to copy this value back to cell E3, updating cell E3. This copy operation is required because in the next loop cycle, cell E4 will again refer to cell E3 for the a_{i-1} value obtained from the previous loop cycle. That way a_n can be recursively calculated from a_1, a_2, a_3, and so on until a_{n-1}.

Copying the value of one cell to another is done using the REP action, defined in cell range G2:K2 (Fig. 3.15b). REP has the following syntax:

```
REP          {destination:r}
             {source:r}
             {num_of_replacements:s=1}
             {op:s=TRUE}
```

which copies the values from one or more source cells to the designated destination cells. This copying operation is done once per REP execution by default, but this can be increased by stating the number of times this copying operation should be done in the num_of_replacements parameter. More will be discussed later about this parameter. The last parameter, op, as usual, denotes if the REP action should be executed (if TRUE or left blank) or cancelled (if FALSE).

Note that the REP action copies only the values from the source cells, not the cell contents which may contain formulas. So, if cell A1 has the value 10 and cell B1 the formula "$=A1+2$", using the REP action to copy B1 to C1 would have the result of the formula in B1 being copied to C1. In other words, cell C1, after the REP action, will contain the value 12.

In Fig. 3.15, cell G2 is defined with the cell name _operation, so that BuildIt is able to execute the REP action to copy the value in cell E4 to E3. This execution is done once every loop cycle, so that by the time the loop ends, cell E3 would have the value for $a5$, as shown in Fig. 3.16.

	A	B	C	D	E	F	G	H	I	J	K
1	INPUT			MODEL			OPERATION				
2	n	5		i	6		REP	123	251		
3				a_{i-1} (or a_n)	123						
4	CONTROL			a_i	251						
5	stepsize	1									
6	step	5									
7	criteria	FALSE									
8											

Fig. 3.16. Result of using the REP action in Excel to recursively determine the value of $a5$ (cell E3).

Typically, num_of_replacements parameter is set to 1 for a single copying operation to occur per REP execution. Nonetheless, this parameter adds flexibility to the REP task.

Consider Eq. 3.5 again. This time we will determine the value of $a5$ without using the loop counter. Fig. 3.17 shows that Eq. 3.5 can still be used to determine $a5$ without using the BuildIt loop.

It is important to manually enter 3 in cell B2 to initialize $a1=3$ (Eq. 3.5). The num_of_replacements parameter (cell G2) of REP is set to 4 to mean four copying operations will be done per REP execution. Why 4? This is because cell B2 already contains the first a_1 value (3). Four more copying operations are needed: first to determine a_2, then a_3, a_4, and finally a_5.

Once implementation is complete, click "Do Operations Only" (not "Start Simulation" because there is no loop) to execute the REP action which gives the result as shown in Fig. 3.18.

	A	B	C	D	E	F	G	H
1	MODEL			OPERATION				
2	a_{i-1} (or a_n)	123		REP	=B2	=B3	4	
3	a_i	=2*B2+5						
4								
5								

Fig. 3.17. Change the REP's num_of_replacements parameter to 4 to determine a_5 (without using BuildIt's loop). Note: $a_i = 2a_{i-1}+5$ and $a_1 = 3$.

	A	B	C	D	E	F	G	H
1	MODEL			OPERATION				
2	a_{i-1} (or a_n)	123		REP	123	251	4	
3	a_i	251						
4								
5								

Fig. 3.18. The value of a_5 (cell B2) determined without using the BuildIt loop by performing four copying operations (cell G2) per REP execution.

One critical disadvantage of this second method is we must remember to enter 3 in cell B2 whenever we run our model. Forgetting to do so will result in erroneous result because of the wrong initial value set for a_1.

To overcome such problems where some variables require initial values at the start of any model run, BuildIt provides the INI action (Table 3.5). Variable initializations using the INI action, however, will be discussed in the next chapter.

3.5.3 Single integration: Canopy photosynthesis

We will continue to build up on our leaf photosynthesis model (first presented in the previous chapter) by determining the canopy photosynthesis, then the daily canopy photosynthesis. Both these examples illustrate the use of the integration action, ITG, within a loop for repetitive calculations.

In the previous chapter, leaf photosynthesis is given by Eq. 2.3 as

$$A_L = \frac{A_m \varepsilon k I_0 \exp(-kL)}{A_m + \varepsilon k I_0 \exp(-kL)}$$

where, as previously given, A_L is the leaf photosynthesis rate (μg CO_2 m^{-2} leaf s^{-1}); A_m is the maximum leaf photosynthesis rate (μg CO_2 m^{-2} leaf s^{-1}); ε is the solar radiation conversion factor (μg CO_2 J^{-1}); k is the canopy extinction coefficient for solar radiation (unitless); I_0 is the solar irradiance above the canopy (W m^{-2} ground or J m^{-2} ground s^{-1}); and L is the cumulative leaf area index from the canopy top to the canopy depth being considered (m^2 leaf m^{-2} ground).

To determine the canopy photosynthesis (A_T; µg CO_2 m^{-2} leaf s^{-1}), Eq. 2.3 has to be integrated over the whole canopy as

$$A_T = \int_0^{LAI} A_L\, dL$$

$$= \int_0^{LAI} \frac{A_m \varepsilon k I_0 \exp(-kL)}{A_m + \varepsilon k I_0 \exp(-kL)} dL \tag{3.6}$$

where LAI is the total leaf area index (m^2 leaf m^{-2} ground).

Note that LAI is the total leaf area of the *whole canopy* per unit ground area, whereas L is the cumulative leaf area (per unit ground area) from the canopy top to a given canopy depth. For instance, at half the total canopy depth, L is the sum of the leaf area from the canopy top to half the total canopy depth. So, as we move progressively down the canopy depth, L increases from 0 to LAI. In short, $0 \leq L \leq LAI$.

In this example, we wish to determine how A_T varies with LAI from 0 to 4. Solving Eq. 3.6 using Excel requires the use of the ITG action for integration and a loop to repeatedly determine A_T at various LAI levels.

Like before, we will continue to use the same values for the equation parameters: A_m as 1500 µg CO_2 m^{-2} s^{-1}; k as 0.5; I_0 as 200 W m^{-2}; and ε as 12 µg CO_2 J^{-1}. All these parameters are treated as constants except for LAI which will vary between 0 to 4 with an interval size of 0.5.

The canopy photosynthesis model is implemented in Excel, as shown in Fig. 3.19, and Table 3.7 shows the cell names defined and used in the model. Fig. 3.20 summarizes the program flow for this canopy photosynthesis model.

a)

	A	B	C	D	E	F	G	H	I
1	CONTROL			MODEL			TO OUTPUT		
2	max steps	4		LAI	=_step			LAI	A_T
3	stepsize	0.5		L				=LAI	=AT
4	step			n1	=Am*e*k*Io*EXP(-k*L)				
5	criteria	=LAI<=B2		n2	=Am+e*k*Io*EXP(-k*L)		OUTPUT		
6				A_L	=E4/E5				
7	PARAMETERS			A_T					
8	Am	1500							
9	k	0.5							
10	Io	200							
11	ε	12							
12									

b)

	I	J	K	L	M	N	O	P	Q
1		OPERATION							
2		ITG	=L	=AL	=AT	0	=LAI		
3									
4									
5									
6									

Fig. 3.19. Implementation of the canopy photosynthesis model in Excel. Setup of the a) loop, calculations, and the output sections, and b) the operation section.

Table 3.7. Cell names defined for the canopy photosynthesis model (BuildIt cell names are listed first).

Cell	Cell name	Cell	Cell name
B3	_stepsize	B9	k
B4	_step	B10	Io
B5	_criteria	B11	e
G3	_read	E2	LAI
G6	_write	E3	L
J2	_operation	E6	AL
B8	Am	E7	AT

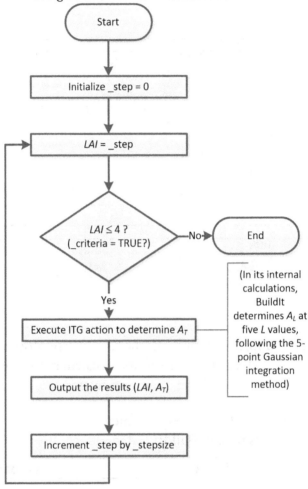

Fig. 3.20. Summary of the program flow for the canopy photosynthesis model.

Cells B8:B11 hold the values for the equation parameters. Cell B2:B5 hold the loop information, where cell B5 (_criteria) in particular instructs BuildIt to end the loop run only when *LAI* (cell E2) is greater than 4 (cell B2). Cells E2:E7 contain the calculations of the model, where cells E4 and E5 are the numerator and denominator of the function in Eq. 3.6, respectively. The result of their division is calculated in cell E6 (A_L).

Integration is performed by the ITG action (cell range J2:Q2). The parameters of this action tell ITG to integrate the function A_L (cell L2 referring to cell E6) from 0 to *LAI* (cells N2 and O2, respectively) and to store the result (cell M2) in cell E7 (A_T).

Cell P2 is left blank because this is a single integral, and cell Q2 is likewise blank to instruct BuildIt to always perform this ITG action. Both cells P2 and Q2 are assumed TRUE by default.

BuildIt finds the cell defined with _read (cell G3) to include *LAI* and *AT* in the model output. This output will be displayed as a list starting from cell G6 (defined with _write).

Once model implementation is completed, click "Start Simulation", and the results are shown in Fig. 3.21, where a chart is additionally drawn using the output listing in cell range G6:H14 to visually depict how *AT* varies with *LAI*.

F	G	H	I	J	K	L	M	N	O	P	Q
1	TO OUTPUT			OPERATION							
2	LAI	A_T		ITG	3.812	159.4	1455	0	4.5		
3	4.5	1455									
4											
5	OUTPUT										
6	0	0									
7	0.5	310									
8	1	577									
9	1.5	802									
10	2	989									
11	2.5	1144									
12	3	1271									
13	3.5	1373									
14	4	1455									

Fig. 3.21. Simulation results by the canopy photosynthesis model.

3.5.4 Double integration: Daily canopy photosynthesis

In the previous example, leaf photosynthesis A_L, or Eq. 2.3, was integrated over the entire canopy, $[0, LAI]$, to give the canopy photosynthesis A_T or Eq. 3.6. This equation, however, only gives the canopy photosynthesis at an instantaneous moment.

To determine the daily canopy photosynthesis (that is, the canopy photosynthesis for the entire day), we must integrate Eq. 3.6 over the whole day, which is equivalent to integrating the function from the time of sunrise to sunset (since there is no photosynthesis for periods before sunrise and after sunset).

The daily canopy photosynthesis equation, $A_{T,d}$, is as shown below, expressed in units g CO_2 day^{-1}:

$$
\begin{aligned}
A_{T,d} &= \frac{3600 \cdot LAI}{10^6} \int_{t_{sr}}^{t_{ss}} A_T \; dt \\[2mm]
&= \frac{3600 \cdot LAI}{10^6} \int_{t_{sr}}^{t_{ss}} \left(\int_0^{LAI} A_L dL \right) dt \qquad (3.7) \\[2mm]
&= \frac{3600 \cdot LAI}{10^6} \int_{t_{sr}}^{t_{ss}} \int_0^{LAI} \frac{A_m \varepsilon k I[\![t]\!] \exp(-kL)}{A_m + \varepsilon k I[\![t]\!] \exp(-kL)} dL \; dt
\end{aligned}
$$

where the division by 10^6 is to convert μg CO_2 to g CO_2, multiplication by 3600 is to convert t, t_{sr} and t_{ss} from unit hours to seconds, and further multiplication by LAI is to convert the photosynthetic rate per unit leaf area to the photosynthetic rate for the whole canopy.

In the previous examples, we had always taken the solar irradiance I_0 as a constant value.

Solar irradiance is actually a function of time, t, where I_0 typically varies sinusoidally with time. Thus, in Eq. 3.7, $I[\![t]\!]$ is the solar irradiance (J m^{-2} ground s^{-1}) at time t (hours), given by France and Thornley (1984) as

$$I_0[\![t]\!] = \frac{I_{t,d}}{1800(t_{ss}-t_{sr})}\sin^2\left[\frac{\pi(t-t_{sr})}{(t_{ss}-t_{sr})}\right] \qquad (3.8)$$

where $I_{t,d}$ is the daily total solar irradiance (J m^{-2} ground day^{-1}); and t_{sr} and t_{ss} are the times of sunrise and sunset, respectively (hours).

We now wish to determine the relationship between $A_{T,d}$ and LAI from LAI = 0 to 4 with a 0.5 interval. We will take the same values for the equation parameters as before, with the following additional new values: $I_{t,d}$ as 10 MJ m^{-2} day^{-1}, t_{sr} as 6.00 hours, and t_{ss} as 18.00 hours.

Fig. 3.22 shows one way the daily canopy photosynthesis model can be implemented in Excel. Cell names defined and used in the model implementation are as listed in Table 3.8.

Like before, cells B2:B5 the hold information about the loop for BuildIt. The loop `_criteria` (cell B5) tells BuildIt to end the loop run when LAI (cell E9) is greater than 4 (cell B2).

Cells B8:B13 store the constant values for the various equation parameters.

Cell E3:E6 are the calculations based on Eq. 3.8 to determine I_0 (cell E6), and cells E9:E13 are the calculations based on Eq. 2.3 to determine A_L (cell E13).

a)

	A	B	C	D	E	F
1	CONTROL			MODEL		
2	max steps	4		*Solar irradiance*		
3	stepsize	0.5		t, hour		
4	step			n1	=ltd/(1800*(tss-tsr))	
5	criteria	=LAI<=B2		n2	=SIN(PI()*(hour-tsr)/(tss-tsr))^2	
6				lo	=E4*E5	
7	PARAMETERS					
8	Am	1500		*Leaf photosyn.*		
9	k	0.5		LAI	=_step	
10	e	12		L		
11	ltd	10000000		n3	=Am*e*k*lo*EXP(-k*L)	
12	tsr	6		n4	=Am+e*k*lo*EXP(-k*L)	
13	tss	18		A_L	=E11/E12	
14						
15				*Daily canopy photosyn.*		
16				ITG result		
17				A_{Td}	=(E16*LAI*3600)/10^6	
18						

b)

	F	G	H	I	J	K	L	M	N	O	P	Q
1		TO OUTPUT			OPERATION							
2		LAI	A_{Td}		ITG	=L	=AL	=E16	0		=LAI	FALSE
3		=LAI	=E17		ITG	=hour	=AL	=E16	=tsr	=tss		
4												
5		OUTPUT										
6												
7												

Fig. 3.22. Implementation of the daily canopy photosynthesis model in Excel. Setup of the a) loop and calculations section, and b) output and operation sections.

Since double integrations are required, two successive ITG actions must be specified in the action list. Cell J2 is defined with the cell name _operation which marks the start of this action list. The first ITG action is to integrate A_L with respect to L and the second ITG with respect to t. Since a double integration is required, the single_integral parameter for the first and second ITG parameter must be set to FALSE (cell P2) and TRUE (or left blank in cell P3), respectively.

Table 3.8. Cell names for the daily canopy photosynthesis model (BuildIt cell names are listed first).

Cell	Cell name
B3	_stepsize
B4	_step
B5	_criteria
G3	_read
G6	_write
J2	_operation
B8	Am
B9	k
B10	e
B11	Itd
B12	tsr
B13	tss
E3	hour
E6	Io
E9	LAI
E10	L
E13	AL
E17	ATd

Result of the double integration is stored in cell E16 which is then multiplied with ($LAI \times 3600/10^6$) in cell E17 to convert the result into units g CO_2 day^{-1} (Eq. 3.7). Cell name _read (cell G3) instructs BuildIt to include *LAI* and $A_{T,d}$ in the model output, which appears as a list starting from cell G6 (defined with _write).

After choosing "Start Simulation" from the BuildIt menu, the model output are as shown in Fig. 3.23. The output listing in cell range G6:H14 is used to plot the chart to show how A_{Td} varies with LAI.

	G	H	I	J	K	L	M	N	O	P	Q
1	TO OUTPUT			OPERATION							
2	LAI	A_{Td}		ITG	3.812	8.851	17192	0	4.5	FALSE	
3	4.5	278.5105		ITG	17.44	8.851	17192	6	18		
4											
5	OUTPUT										
6	0	0									
7	0.5	6.231946									
8	1	23.39698									
9	1.5	49.28468									
10	2	81.86393									
11	2.5	119.3463									
12	3	160.2233									
13	3.5	203.2767									
14	4	247.5649									
15											
16											
17											
18											

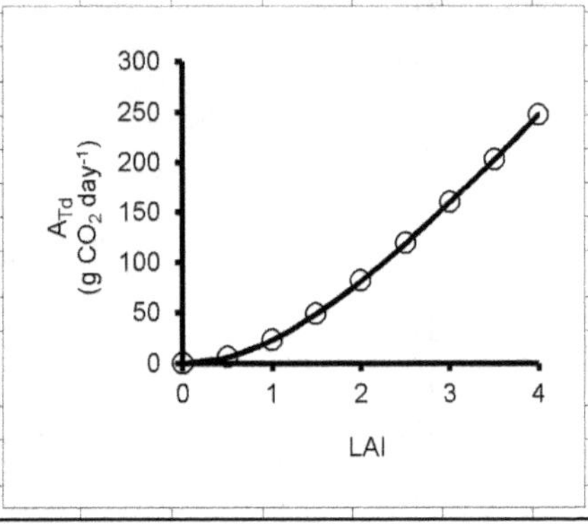

Fig. 3.23. Simulation results by the daily canopy photosynthesis model.

Three other BuildIt actions: INI, UPD, and RUN will be discussed in the next chapter.

3.6 Exercises

1. Use BuildIt actions DIF and ITG to solve the following:

   a) $y = 2x\exp(-x)$
   Determine y for $x = 2$.

   b) $y = -\cos(x^2)$
   Determine y for $x = 0$ to 5 at every 0.5 unit interval.

c) $z = \int_{0}^{2} x \sin\left(x^2\right) dx$ where x is in unit radian.

Determine z.

d) $z = \int_{0}^{3}\int_{0}^{2} (x-n)^2 + \sin\left(y/n\right) dx dy$

Determine z for $n = 1$ to 5 at every 1 unit interval.

2. BuildIt has several actions for list operations. Use one or more of them in Excel to obtain the following results:

a) Given a list of numbers (40, 9, 33, 8, 2), sort ascendingly the values in this list then shift each number to the left twice to produce the following result: (9, 33, 40, 2, 8).

b) Given two lists of numbers: (45, 0, 20) and (3, 19, 1), sort descendingly both these lists then add their corresponding numbers; that is, after sorting the two lists, and they are added together to give: (45, 20, 0) + (19, 3, 1) = (64, 23, 1).

c) Randomize the order of numbers in the following list (10, 20, 30, 40, 50) and multiply then members in this the list by 2.
After randomization, the numbers in the list may have the following order: (20, 30, 10, 50, 40), after which each of these numbers is multiplied by 2 to give (40, 30, 20, 100, 80).

d) Without using Excel's MIN and MAX functions, how do we use BuildIt to determine the smallest and largest value in a given list?

For instance, given a list (3, 18, 0, 2, 14), use BuildIt to determine that 0 and 18 are the minimum and maximum values, respectively.

3. The total number of combinations r elements that can be selected from N total elements is determined by

$$C = \frac{N!}{r!(N-r)!}$$

where C is the total number of combinations; N is the total number of elements; and r is the number of elements selected or chosen.

For instance, how many possible ways can we select 3 (r) students out a class of 25 (N) students? Using the above equation, we get

$$C = \frac{25!}{3!(25-3)!} = 2,300 \text{ combinations.}$$

Use BuildIt's ACC action to implement this equation for any N and r values.

Hint: ACC's last parameter, op, should have a logical condition to check if the cumulative multiplication should be carried out.

4. Use the REP action to implement the following equation in Excel:

$$x_i = 1 + \sqrt{x_{i-1}}$$

where $i > 0$ and $x_0 = 2$. Determine x_{10}.

5. From Eq. 3.6, use BuildIt to determine A_T for $I_0 = 0$ to 800 W m^{-2} at every 100 W m^{-2} interval. Set A_m as 1500

μg CO_2 m^{-2} s^{-1}, k as 0.5, ε as 12 μg CO_2 J^{-1}, and LAI as 2 m^{-2} leaf m^{-2} ground.

Using the model output, plot A_T against I_0 values.

6. From Eq. 3.7 and 3.8, determine $A_{T,d}$ for $I_{t,d}$ = 10^6 to 10^7 J m^{-2} day^{-1} at every 10^6 J m^{-2} day^{-1} interval. Set A_m as 1500 μg CO_2 m^{-2} s^{-1}, k as 0.5, ε as 12 μg CO_2 J^{-1}, LAI as 2 m^{-2} leaf m^{-2} ground; and t_{sr} and t_{ss} as 6 and 18 hours, respectively.

Using the model output, plot $A_{T,d}$ against $I_{t,d}$ values.

7. Environmental stress can reduce a plant's photosynthetic rate according to

$$\hat{A}_{T,d} = A_{T,d} \times R_D$$

where $A_{T,d}$ is the daily potential (maximum possible for the day) photosynthetic rate which is reduced by environmental stress R_D to $\hat{A}_{T,d}$. R_D is a stress level ranging from 0 (maximum stress) to 1 (no stress), and it is computed simply by

$$R_D = \begin{cases} 1 - \left(\dfrac{20 - T_a}{10} \right) & 10 \leq T_a \leq 20 \text{ °C} \\ 1 & 20 < T_a < 30 \text{ °C} \\ 1 - \left(\dfrac{T_a - 30}{10} \right) & 30 \leq T_a \leq 40 \text{ °C} \\ 0 & T_a < 10 \text{ °C or } T_a > 40 \text{ °C} \end{cases}$$

where T_a is the ambient air temperature (°C). In this case, the stress on the plant is strictly due to air temperatures that are too low or too high.

Using Eq. 3.7 and 3.8, determine $\hat{A}_{T,d}$ for T_a = 5 to 45 °C at every 5 °C interval. Set A_m as 1500 µg CO_2 m^{-2} s^{-1}, k as 0.5, ε as 12 µg CO_2 J^{-1}, $I_{t,d}$ = 10^7 J m^{-2} day^{-1}, LAI as 2 m^{-2} leaf m^{-2} ground; and t_{sr} and t_{ss} as 6 and 18 hours, respectively.

Chapter 4. BuildIt actions (Part 2)

4.1 Modeling with differential equations

Equations that involve rates of change (derivatives) are called differential equations. Differential equations describe how quickly variables change, and these equations are very common in crop models because they are often used to describe the changes (per unit time) in properties such as plant weight, leaf area, plant height, soil water content, and pest population.

You can solve differential equations using the DIF action as discussed in the previous chapter, but using the UPD action allows us to estimate future values of variables based on their current values and rates of change.

For instance, the future value of variable x can be estimated, using Euler's method, by

$$x_{t+\Delta t} = x_t + \frac{dx_t}{dt} \Delta t \qquad (4.1)$$

where x_t and $x_{t+\Delta t}$ are the values of x at time steps t and $t+\Delta t$, respectively; Δt is the interval size between the two successive time steps; and the differential equation dx_t/dt is the rate of change of x at time t.

Eq. 4.1 however makes an important assumption that, within the Δt period, the rate of change dx_t/dt would remain constant, so that the future value of $x_{t+\Delta t}$ can be estimated simply by multiplying its current value (x_t) with its current rate of change (Fig. 4.1). Consequently, the interval size Δt is important because if this interval size is too large, this can cause the estimation of $x_{t+\Delta t}$ to be unacceptably inaccurate in particular when the dx_t/dt changes rapidly within the Δt period.

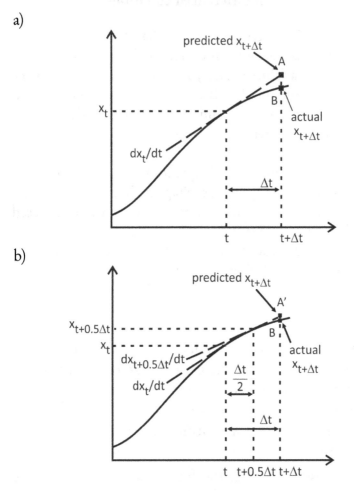

Fig. 4.1. Prediction error of $x_{t+\Delta t}$: a) can be large when the growth rate, dx_t/dt, is assumed constant over a large interval Δt, but b) the error can be reduced when Δt is divided into two (or more) subintervals.

Fig. 4.1a shows that at time step t, the value of x is x_t and the rate of change is dx_t/dt. So, in the next time step $t+\Delta t$, the value for $x_{t+\Delta t}$, calculated from Eq. 4.1, is denoted by point A in the chart.

Reaching point A, however, requires the assumption that within the time interval Δt, the rate of change in x_t is constant, but Fig. 4.1a clearly shows that this is untrue. The rate of change actually decreased within this period, so the actual $x_{t+\Delta t}$ value is really at point B. So, the larger the Δt, the higher the risk that the predicted value may deviate far from the true value (*i.e.*, distance between A and B is large).

To reduce the estimation error, we can divide Δt into two or more smaller subintervals, and for each successive subinterval period, we use Eq. 4.1 to update the value of x. In Fig. 4.1b, Δt is divided into two equal subintervals, so that $x_{t+\Delta t}$ (point A') is estimated in two steps:

$$x_{t+0.5\Delta t} = x_t + \frac{dx_t}{dt} \cdot 0.5\Delta t$$

in the first half of the Δt period, followed by

$$x_{t+\Delta t} = x_{t+0.5\Delta t} + \frac{dx_{t+0.5\Delta t}}{dt} \cdot 0.5\Delta t$$

in the second half.

Dividing Δt into even more subintervals would further reduce the estimation error but may lead to excessive number of computations and time-consuming model runs with little gains in estimation accuracy. Consequently, some trial-and-error experimentation is required to determine the optimal number of subintervals to have within Δt.

The UPD BuildIt action uses the Euler's method (Eq. 4.1) ιto estimate the future value of a given variable by multiplying its current value by its current rate of change. This action has the following syntax:

```
UPD        {x:r}
           {rate_of_change:r}
           {n:s=1}
           {op:s=TRUE}
```

which updates variable x by Eq. 4.1, where rate_of_change is the differential equation dx_t/dt, and Δt is the interval size (specified by cell name _stepsize). The parameter n (defaults to 1 if n is unspecified) is the number of subintervals to have within each Δt period, and last parameter op specifies that the UPD action should be executed if op is set to TRUE (or left blank) or to cancel if op set to FALSE.

The UPD action typically requires the INI action to initialize the variable x at the start of a model run. The INI action has the following syntax:

```
INI        {destination:r}
           {source:r}
           {op:s=TRUE}
```

which copies the cell values from source to destination. INI action is used primarily for setting initial values to model parameters. Like the REP action, the INI action only copies the cell values (not contents), so if a source cell contains a formula, the result of the formula will be copied to the destination cell.

Consider the following differential equation:

$$\frac{dx_t}{dt} = 2x \tag{4.2}$$

We shall fix Δt as 1 and the initial value of x at $t=0$ as 1 (*i.e.*, $x0=1$). So, at $t=1$, $x1$ is determined using Eq. 4.1 and 4.2 as:

$$x_1 = (x_0 + 2x_0 \cdot 1) = (1+2) = 3$$

In the same way, $x2$ (at $t=2$) is determined as:

$$x_2 = (x_1 + 2x_1 \cdot 1) = (3+6) = 9$$

and so on. Note that since Eq. 4.2 is a linear line, dividing Δt into more than one subinterval is unnecessary as having more subintervals in this case will not increase the estimation accuracy of x.

We can now see that we require a loop and the UPD action to iteratively determine x at various time steps, and we also need the INI action to initialize x so that $x0$ will always have the desired initial value at the start of every model run. Lastly, we need to output the (t, x_t) pair of values.

The main steps to implement Eq. 4.1 and 4.2 in Excel are as follows:

a) Create the loop section which contains information for BuildIt to maintain and run the loop (*e.g.*, define cell names _step, _stepsize, and _criteria).

b) Let one cell hold the value of the x variable and another cell to calculate its rate of change based on Eq. 4.2.

c) Create the operation section by defining a cell with the BuildIt cell name _operation. This cell marks the start of the action list BuildIt will read and execute at every loop cycle.

In the action list, specify the UPD action which would use Eq. 4.1 to determine the various values of x at successive time steps.

d) Create the prerun section by defining a cell with the BuildIt cell name _prerun (Fig. 4.2).

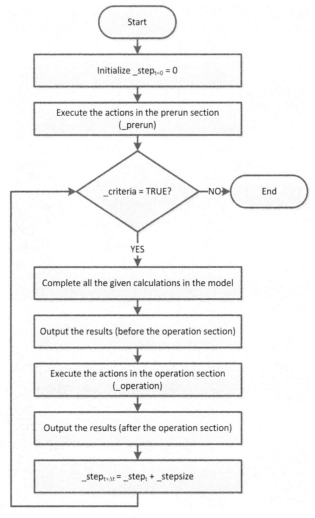

Fig. 4.2. Program flow of BuildIt which includes the prerun section and the option to output the model results before or after the operation section.

Similar to the operation section, the cell defined with `_prerun` name also marks the start of the action list BuildIt will read and execute. However, unlike those in the operation section, the actions listed in the prerun section are only read and executed *once* by BuildIt. Their executions also occur at the start of a simulation or model run and before the loop begins.

The prerun section is typically used to initialize selected variables. In this example, we need to initialize x with 1 (*i.e.*, x_0 = 1), so the INI action should be specified in the prerun section for x initialization.

e) Lastly, define one cell with the cell name `_read` and another cell with `_write` to specify what parameters to include in the model output list and where this list should be stored, respectively. However, unlike what you have been shown before, BuildIt actually has two occasions to store the output: once before and once after the actions in the operation section are executed (Fig. 4.2).

Fig. 4.3 is an example of a what-to-output section. Cell K3 is defined with the cell name `_read`, so the cell range starting from K3 until O3 (stops at O3 because the next cell, P3, is blank) instruct BuildIt to include the `_step` loop counter and the values in cells E1, E2, and E3 in the model output.

However, notice the cell range below K3:O3. This cell range K4:O4 instruct BuildIt whether the corresponding what-to-output parameter should be outputted before (if set to FALSE) or after (if set to TRUE or left blank) the operation section. In Fig. 4.3, `_step` and cells E1 and E3 are to be outputted after the operation section (since cells

89

K4 , L4, and O4 are either blank or set to TRUE) and cells E2 and E3 after the operation section (since cells M4 and N4 are set to FALSE).

	K	L	M	N	O	P
1	TO OUTPUT					
2	t	x	y	z (before)	z (after)	
3	=_step	=E1	=E2	=E3	=E3	
4		TRUE	FALSE	FALSE	TRUE	
5						

Fig. 4.3. You can specify in the what-to-output section whether a given parameter or variable should be outputted before or after the operation section. Set TRUE (or leave blank) if the corresponding what-to-output parameter is to be outputted after the operation section, or else set to FALSE for output to occur before the operation section.

Notice in Fig. 4.3 that cell E3 (for variable z) is to be outputted twice: once before (cell N4 is set to FALSE) and again after (cell O4 is set to TRUE) the operation section. In most cases, it does not matter if a variable or parameter is to be outputted before or after the operation section because their before or after values would be the same as each other. But this is not always the case. If any actions in the operation section alter the value of a variable, the variable's value before and after the operation section will likely not be the same as each other.

The UPD action, for instance, takes x_t and multiplies it with its current rate of change and Δt to update x_t to $x_{t+\Delta t}$. Consequently, the values of x_t and $x_{t+\Delta t}$ would likely not the same as each other. In this case, the variable x should be outputted before the operation section, so there is a match between the current value of x and the time step t. In other words, the output of x before the operation section would produce the (t, x_t) matching pair. But if the output of x was done after the operation section, the output would be the mismatch pair of $(t, x_{t+\Delta t})$.

To better understand how UPD, INI, and other actions are used, let us now consider three examples of how differential equations can be used and implemented in Excel, starting with a trivial example, then followed by a prey-predator model, and lastly, a simple crop growth model.

4.2 Simple differential equation

The rate of change of y is given by:

$$\frac{dy_t}{dt} = 2 - \exp(-4t) - 2y \qquad (4.3)$$

where y_t is the value of y at time t; and exp(-4t) denotes the natural exponential function e^{-4t}. The value of $y_{t+\Delta t}$ at the subsequent time step $t + \Delta t$ is estimated based on Eq. 4.1 as

$$y_{t+\Delta t} = y_t + \frac{dy_t}{dt}\Delta t \qquad (4.4)$$

where Δt is the interval size between the time steps t and $t + \Delta t$.

Let us fix the initial y value at time $t=0$ as 1, and we wish to determine y at $t=0.1, 0.2, ..., 0.5$.

The cell names used and defined in the worksheet to implement Eq. 4.3 and 4.4 are listed in Table 4.1, and the model implementation is shown in Fig. 4.4.

Cells B2:B5 contain the loop information where Δt is represented by _stepsize (cell B3) set to 0.1. Cell B3 (_stepsize) and B5 (_criteria) instruct BuildIt to run the loop for t = 0, 0.1, 0.2, and so on until 0.5 at 0.1 time interval.

Table 4.1. Cell names defined in the worksheet to implement the simple differential equation $2-\exp(-4t)-2y$ to estimate y_t at various time steps (BuildIt cell names are listed first).

Cell	Cell name	Cell	Cell name
B3	_stepsize	E2	time
B4	_step	E3	y
B5	_criteria	E4	dy_dt
G2	_operation		
G6	_prerun		
G11	_read		
G15	_write		

The current time step t is cell E2 (defined with cell name `time`) which refers to the loop counter `_step` (cell B4). Cell E3 holds the current y value (y_t). The cell is blank because UPD action will use this cell in its internal calculations to estimate y_{t+1} based on Eq. 4.3 which is implemented in cell E4 (defined with cell name `dy_dt`). In cell E4, Excel's EXP natural exponential function is used to determine e^{-4t} (*see* Eq. 4.3).

Cell G6 is defined with the cell name `_prerun`, so cell G6 marks the start of the actions to execute before the loop cycle begins (Fig. 4.2). There is only one action specified: INI. This action would ensure y (cell E3) will always start with the value of 1 (cell I6) before any calculations in the model begin. Recall the actions in the prerun section are executed only once unlike those in the operation section which are executed once every loop cycle.

a)

	A	B	C	D	E	F
1	CONTROL			MODEL		
2	maxsteps	0.5		t	=_step	
3	stepsize	0.1		y		
4	step			dy/dt	=2-EXP(-4*time)-2*y	
5	criteria	=time<=B2				
6						

b)

	F	G	H	I	J	K
1		OPERATION				
2		UPD	=y	=dy_dt	5	
3						
4						
5		PRERUN				
6		INI	=y	1		
7						
8						
9		TO OUTPUT				
10		t	y			
11		=time	=y			
12			FALSE			
13						
14		OUTPUT				
15						
16						

Fig. 4.4. Implementation of the differential equation $2-\exp(-4t)-2y$ in Excel. Setting up of: a) the loop and calculation sections, and b) the what-to-output and output sections, as well as specifying the BuildIt actions to be executed in the operation and prerun sections.

Cell G2 is defined with the cell name _operation, so this cell marks the start of the operation section which, in this case, contains only the UPD action. This action is used iteratively within every loop cycle to estimate $y_{t+\Delta t}$ based on Eq. 4.4.

Notice that cell J2 (representing UPD's parameter n) is arbitrarily set to 5 to mean Δt is divided into 5 subintervals. You can set cell J2 to other values (*e.g.*, compare having one with five subintervals) to determine how the number of subintervals would affect the estimations of y.

Cell G11 is defined with the cell name _read to instruct BuildIt to include the pair of (t, y_t) in the model output. Look at cells G12 and H12 in particular. Cell G12 is blank so it is TRUE by default. This setting tells BuildIt to output t after BuildIt executes any action it finds in the operation section. In contrast, cell H12 is FALSE to instruct BuildIt to output y_t before BuildIt reads the operation section. It is important to output y_t before (not after) the operation section so that the output of y_t matches the current time t. For the case of t, it does not matter if cell G12 is set to TRUE or FALSE since t is not altered within the operation section.

Lastly, cell G15 is defined with the cell name _write so that the model output of (t, y_t) will be displayed starting from cell G15 onwards.

Choose the "Start Simulation" from the BuildIt menu to run the model, and the result is as shown in Fig. 4.5a. A chart is additionally drawn from the (t, y_t) output listing (cell range G15:H20) to visually depict how y varies with time t.

What happens if we set cell H12 to TRUE instead of FALSE so that the output of y occurs only after the operation section?

a)

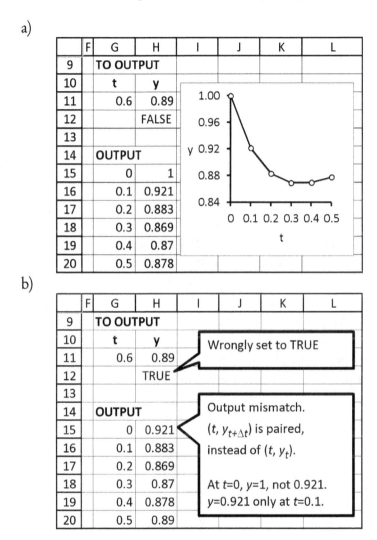

b)

	F	G	H	I	J	K	L
9		TO OUTPUT					
10		t	y				
11		0.6	0.89				
12			TRUE				
13							
14		OUTPUT					
15		0	0.921				
16		0.1	0.883				
17		0.2	0.869				
18		0.3	0.87				
19		0.4	0.878				
20		0.5	0.89				

Wrongly set to TRUE

Output mismatch. $(t, y_{t+\Delta t})$ is paired, instead of (t, y_t).

At $t=0$, $y=1$, not 0.921. $y=0.921$ only at $t=0.1$.

Fig. 4.5. Model output: a) estimations of y_t at $t = 0, 0.1, ..., 0.5$ using the differential equation $2-\exp(-4t)-2y$, and b) a mismatch in pairing between time t and y if the latter is to be outputted after the UPD action is executed in the operation section.

In this case, the output is as shown in Fig. 4.5b. We can see there is a mismatch between y and t where the output is actually (t, y_{t+1}), not (t, y_t) as it should in Fig. 4.5a.

4.3 Prey-predator model

Take an ecosystem with only two species: a prey and a predator.

Their populations are closely related to each other. A large population of prey, for instance, means more food is available to the predator. This encourages predation, and this will increase the population of predator.

Eventually predation outstrips food supply which in turn causes a decline in the predator population. A low population of predator now encourages the rise again in the prey population.

This dynamics continue as a cycle of rise and fall in both the prey and predator population.

The prey and predator populations at any one time can be estimated using Euler's method as

$$x_{t+\Delta t} = x_t + \frac{dx_t}{dt}\Delta t \qquad (4.5a)$$

$$y_{t+\Delta t} = y_t + \frac{dy_t}{dt}\Delta t \qquad (4.5b)$$

where x and y are the populations for the prey and predator, respectively; subscripts t and $t+\Delta t$ indicate the population at time steps t and $t+\Delta t$, respectively; Δt is the interval size between the two successive time steps; and the differential equations dx_t/dt and dy_t/dt are the rates of population growth for the prey and predator at time t, respectively.

In the classical Lotka-Volterra's prey-predator model, the rates of population growth over time for the prey (x) and predator (y) can be described by

$$\frac{dx_t}{dt} = \alpha x_t - \beta x_t y_t \tag{4.6a}$$

$$\frac{dy_t}{dt} = \delta x_t y_t - \gamma y_t \tag{4.6b}$$

where α and δ are the growth rates of the prey and predator, respectively; and β and γ are the death rates of the prey and predator, respectively.

Eq. 4.6 shows how closely the two populations of prey and predator are related to each other.

In this example, we are interested to simulate the prey and predator population dynamics for 200 time steps, with a time step interval of 1 ($\Delta t = 1$). In other words, we want to determine the prey and predator populations (x_t, y_t) at time $t = 1, 2, 3, \dots 200$.

Let us fix the initial values for x and y as 50 and 15, respectively; α and δ as 0.1 and 0.001, respectively; and β and γ as 0.01 and 0.05, respectively.

Fig. 4.6 show one way how can we implement Eq. 4.5 and 4.6 in Excel.

Because of the large model output (output for 200 time steps), it is recommended that we create a second worksheet in the workbook specifically just to store the model output (Fig. 4.7).

You can name these two worksheets to whatever names you prefer, but in this book, the first worksheet is named "Model" and the second "Output".

Table 4.2 lists the cell names used in the model.

a)

	A	B	C
1	CONTROL		
2	maxsteps	200	
3	stepsize	1	
4	step		
5	criteria	=_step<=B2	
6			
7	PARAMETERS		
8	*Initial pop.*		
9	prey	50	
10	predator	15	
11	*Growth rate*		
12	prey	0.1	
13	predator	0.001	
14	*Death rate*		
15	prey	0.01	
16	predator	0.05	
17			

b)

	C	D	E	F	G
1		MODEL			
2			pop.	rate of change	
3		Prey		=prey_growth*prey_pop-prey_death*prey_pop*pred_pop	
4		Predator		=pred_growth*prey_pop*pred_pop-pred_death*pred_pop	
5					
6					
7					

c)

	G	H	I	J	K	L
1		OPERATION				
2		UPD	=E3:E4	=F3:F4	5	
3						
4						
5						
6						
7						
8		PRERUN				
9		INI	=E3:E4	=B9:B10		
10						

Fig. 4.6. Model worksheet: implementation of the prey-predator model: a) information for the loop and values for the equation parameters, b) setup of the model calculations, and c) setup of the operation and prerun sections.

Table 4.2. Cell names defined and used in the "Model" and "Output" worksheets to implement the prey-predator model (BuildIt cell names are listed first).

Worksheet	Cell	Cell name
Model	B3	_stepsize
	B4	_step
	B5	_criteria
	H2	_operation
	H9	_prerun
Output	A3	_read
	A7	_write
Model	B12	prey_growth
	B13	pred_growth
	B15	prey_death
	B16	pred_death
	E3	prey_pop
	E4	pred_pop
	F3	prey_rate
	F4	pred_rate

	A	B	C	D
1	TO OUTPUT			
2	t	prey pop.	predator pop.	
3	=_step	=prey_pop	=pred_pop	
4		FALSE	FALSE	
5				
6	OUTPUT			
7				
8				
9				

Fig. 4.7. Output worksheet: due to the large model output, the output is located in a separate worksheet named "Output". Here shows the setup of the what-to-output and output sections. The output will be a list of (t, x_t, y_t).

In Fig. 4.6, cells B2:B5 contain the loop information where we can see that the loop will run for $t = 0$ to 200 with a time interval size of 1.

Cells B9 and B10 hold the initial values for the prey and predator populations, respectively, and cells B12:B13 and B15:B16 the constants for the prey and predator growth and death rates.

Cells F3 and F4 implement Eq. 4.6a and b, respectively, and cells E3 and E4 hold the current populations of the prey and predator, respectively.

Cell H9 is defined with the cell name _prerun to mark the start of the prerun section. This section contains the INI action to ensure values from cells B9 and B10 are copied to cells E3 and E4, respectively. In other words, the populations of prey (cell E3) and predator (cell E4) at the start of every model run are always initialized with 50 (cell B9) and 15 (cell B10), respectively.

Cell H2 is defined with the cell name _operation to mark the start of the operation section. The single UPD action estimates the prey and predator populations at time $t + 1$ based on their current (at time t) population and rates of change.

Cell K2 specifies that the UPD's n parameter is set to 5 to mean the time interval Δt is to be divided into five subintervals to increase the estimation accuracy of the prey and predator populations.

Note that, instead of a single INI action for simultaneous initializations (Fig. 4.6), we can have two INI actions: one to initialize the prey population, then followed by the second for the predator population (Fig. 4.8).

	G	H	I	J	K	L
1		OPERATION				
2		UPD	=E3:E4	=F3:F4	5	
3						
4						
5						
6						
7						
8		PRERUN				
9		INI	=E3	=B9		
10		INI	=E4	=B10		
11						

Fig. 4.8. We can alternatively use two INI actions instead of one to separately initialize the prey and predator populations.

But unlike for the INI action, we must specify only a single UPD action to estimate the prey and predator populations. So, what is the difference between using a single UPD action and two UPD actions (Fig. 4.6c vs. Fig. 4.9)?

	G	H	I	J	K	L	M
1		OPERATION					
2		UPD	=E3	=F3	5	*No, we*	
3		UPD	=E4	=F4	5	*cannot have*	
4						*two UPD*	
5						*actions in this*	
6						*model.*	
7							
8		PRERUN					
9		INI	=E3:E4	=B9:B10			
10							

Fig. 4.9. Asynchronous cell updates: unlike for the INI action, we cannot have two UPD actions: the first UPD to estimate the prey population, then the second for the predator population. This is because their rates of change depend on each other's current population (at time t).

The difference between these two cases is rather subtle. Eq. 4.6a and b reveal that the rates of change in the prey and predator populations depend on each other; that is, cells F3 and F4 simultaneously use information from cells E3 (prey population) and E4 (predator population) (Fig. 4.6b).

This interdependency means we cannot have two separate and independent updates on the prey and predator populations. In Fig. 4.9, the first UPD action updates the prey population from x_t to x_{t+1} using Eq. 4.5a and 4.6a. So, by the time of the second UPD action, it is x_{t+1} not x_t that is used in Eq. 4.6b to update the predator population from y_t to y_{t+1}.

In the second worksheet, named "Output", cells A3 and A7 are defined with cell names _read and _write, respectively (Fig. 4.7).

Output will include the time and the corresponding prey and predator populations, or (t, x_t, y_t). This output will be displayed starting at cell A7.

Note that cells B4 and C4 are set to FALSE. This is to ensure that the prey (cell B3) and predator (cell C3) populations are to be outputted before the operation section.

Recall the operation section contains the UPD action which would update the prey and predator populations at time t to that at time $t+1$. Setting cells B4 and C4 to TRUE (or leaving them blank) would output the prey and predator populations after the operation section.

In this case, the output listing would be (t, x_{t+1}, y_{t+1}), not the intended (t, x_t, y_t). For the time t, it does not matter if it is outputted before or after the operation as its value would not change within the operation section.

Once model implementation is completed, click "Start Simulation" from the BuildIt menu to start the simulation of the prey-predator population dynamics.

The results are as shown in Fig. 4.10.

The charts drawn from the output listing show that the prey and predator populations would vary widely with time, in particular the prey population.

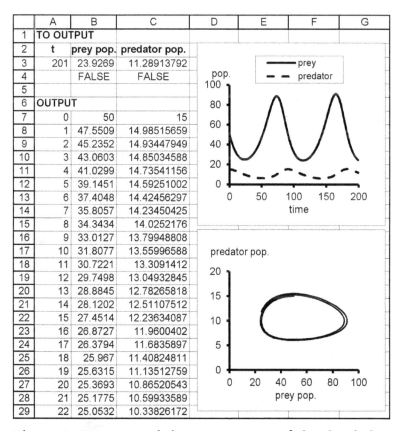

	A	B	C	D	E	F	G
1	TO OUTPUT						
2	t	prey pop.	predator pop.				
3	201	23.9269	11.28913792				
4		FALSE	FALSE				
5							
6	OUTPUT						
7	0	50	15				
8	1	47.5509	14.98515659				
9	2	45.2352	14.93447949				
10	3	43.0603	14.85034588				
11	4	41.0299	14.73541156				
12	5	39.1451	14.59251002				
13	6	37.4048	14.42456297				
14	7	35.8057	14.23450425				
15	8	34.3434	14.0252176				
16	9	33.0127	13.79948808				
17	10	31.8077	13.55996588				
18	11	30.7221	13.3091412				
19	12	29.7498	13.04932845				
20	13	28.8845	12.78265818				
21	14	28.1202	12.51107512				
22	15	27.4514	12.23634087				
23	16	26.8727	11.9600402				
24	17	26.3794	11.6835897				
25	18	25.967	11.40824811				
26	19	25.6315	11.13512759				
27	20	25.3693	10.86520543				
28	21	25.1775	10.59933589				
29	22	25.0532	10.33826172				

Fig. 4.10. Output worksheet: an excerpt of the simulation results from the prey-predator model.

4.4 Simple crop growth model

In this third example, we will implement a simple crop growth model adapted from Thornley (1977) (Fig. 4.11).

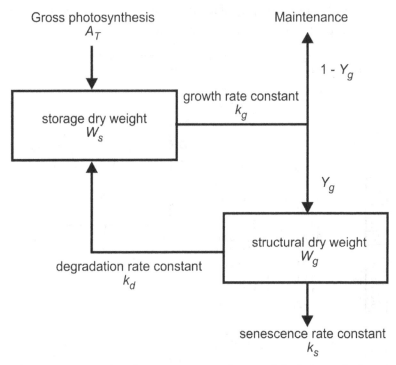

Fig. 4.11. A simple crop growth model (adapted from Thornley, 1977).

The assimilates produced by photosynthesis is partitioned to maintenance for the plant's continual survival and to growth for the synthesis of new materials for the plant. The plant weight consists of two components: the storage dry weight (W_s; g C m^{-2} ground) and the plant structural dry weight (W_g; g C m^{-2} ground).

The storage dry weight W_s is supported by the addition of newly produced substrates via the photosynthesis process.

A portion of the plant's storage, $k_g W_s$, will be utilized for maintenance and growth, where of this total ($k_g W_s$), Y_g of it will be utilized for growth, and the remainder ($1 - Y_g$) for maintenance.

Of the total plant structural weight W_g, k_d of it will degrade and the substrates returned to the storage, and k_s of it will senesce (*i.e.*, loss of materials due to increasing age).

The rates of change in the dry weights for storage and structure components at time t (in unit hours) are thus:

$$\frac{dW_s}{dt} = k_d W_g - k_g W_s + \frac{12}{44} \cdot \frac{3600 \cdot LAI}{10^6} \cdot A_T \qquad (4.7a)$$

$$\frac{dW_g}{dt} = k_g W_s Y_g - k_d W_g - k_s W_g \qquad (4.7b)$$

where A_T is the canopy photosynthesis (μg CO_2 m^{-2} leaf s^{-1}) as given by Eq. 3.6 in the previous chapter.

The other equation parameters and their respective values to be used in this example are listed in Table 4.3.

Note that we multiplied A_T in Eq. 4.7a with $(12/44) \times (3600 \times LAI/10^6)$ to convert A_T's original units from μg CO_2 m^{-2} leaf s^{-1} to g C m^{-2} ground hour^{-1}. Recall that the atomic weight of C and molecular weight of CO_2 are 12 and 44 g, respectively; thus, every 44 g of CO_2 has 12 g of C. Finally, division by 10^6 (1 g = 10^6 μg) and multiplication by 3600 (1 hour = 60 min hour^{-1} × 60 s min^{-1}) and LAI (1 m^2 leaf per 1 m^{-2} ground) is to convert A_T into units g C m^{-2} ground hour^{-1}.

Table 4.3. Parameter values for the simple crop growth model.

Symbol	Description	Units	Value
A_m	maximum leaf photosynthesis rate	$\mu g\ CO_2\ m^{-2}$ leaf s^{-1}	1500
ε	solar radiation conversion factor	$\mu g\ CO_2\ J^{-1}$	12
k	canopy extinction coefficient	-	0.5
LAI	leaf area index	m^2 leaf m^{-2} ground	2
k_g	growth rate constant	day^{-1}	1.98
k_d	degradation rate constant	day^{-1}	0.10
k_s	senescence rate constant	day^{-1}	0.05
Y_g	efficiency of structural synthesis	-	0.75
W_s	substrates dry weight	$g\ C\ m^{-2}$ ground	20 (initial)
W_g	structural dry weight	$g\ C\ m^{-2}$ ground	130 (initial)
t_{sr}	time of sunrise	hour	6
t_{ss}	time of sunset	hour	18
$I_{t,d}$	daily total solar irradiance	$J\ m^{-2}$ ground day^{-1}	10^7

The increase in the dry weights for storage and structure from time t_1 to t_2 are obtained by integrating Eq. 4.7a and b over the period $[t_1, t_2]$ to give

$$\Delta W_s = \int_{t_1}^{t_2} \left(k_d W_g - k_g W_s + \frac{12}{44} \cdot \frac{3600 \cdot LAI}{10^6} \cdot A_T \right) dt$$

$$\hspace{3cm} (4.8a)$$

$$= \left(k_d W_g - k_g W_s \right) \Delta t + \frac{12}{44} \cdot \frac{3600 \cdot LAI}{10^6} \int_{t_1}^{t_2} A_T \, dt$$

$$\Delta W_g = \int_{t_1}^{t_2} \left(k_g W_s Y_g - k_d W_g - k_s W_g \right) dt$$

$$\hspace{3cm} (4.8b)$$

$$= \left(k_g W_s Y_g - k_d W_g - k_s W \right) \Delta t$$

where Δt is the time interval ($t_2 - t_1$), and k_g, k_s, and Y_g are taken as constants.

If we take $\Delta t = 24$ hours or 1 day for daily time steps, then Eq. 4.8a and b simplify to

$$\Delta W_s = \left(k_d W_g - k_g W_s \right) + \frac{12}{44} \cdot \frac{3600 \cdot LAI}{10^6} \int_{t_1=t_{sr}}^{t_2=t_{ss}} A_T \, dt \quad (4.9a)$$

$$\Delta W_g = k_g W_s Y_g - k_d W_g - k_s W_g \hspace{2cm} (4.9b)$$

Since no photosynthesis occurs for periods before sunrise and after sunset, integration of A_T over one day is equivalent to its integration over the period between the hours of sunrise (t_{sr}) and sunset (t_{ss}).

Calculations in Eq. 4.9 are straightforward with the exception of the integration of A_T. Recall from Eq. 2.3 and 3.6 that canopy photosynthesis is given by

$$A_T = \int_0^{LAI} A_L \, dL$$

$$= \int_0^{LAI} \frac{A_m \varepsilon k I_0 \llbracket t \rrbracket \exp(-kL)}{A_m + \varepsilon k I_0 \llbracket t \rrbracket \exp(-kL)} dL$$

where solar irradiance I_0 is not a constant but varies with time t according to Eq. 3.8 as

$$I_0\llbracket t \rrbracket = \frac{I_{t,d}}{1800\left(t_{ss} - t_{sr}\right)}\sin^2\left[\frac{\pi\left(t - t_{sr}\right)}{\left(t_{ss} - t_{sr}\right)}\right] \qquad t_{sr} \leq t \leq t_{ss}$$

where $I_{t,d}$ is the daily total solar irradiance (J m^{-2} ground day^{-1}).

The dry weights for storage and structure can be determined by:

$$W_{s,t_2} = W_{s,t_1} + \Delta W_s \qquad\qquad (4.10a)$$

$$W_{g,t_2} = W_{g,t_1} + \Delta W_g \qquad\qquad (4.10b)$$

where, for their initial weights, we shall fix W_s and W_g as 20 and 130 g C m^{-2} ground, respectively (Table 4.3).

Fig. 4.12 to 4.15 show the implementation of the simple crop growth in Excel, and Table 4.4 lists the cell names defined and used in the worksheet.

In Fig. 4.12, cells B8:B18 and B21:B22 store the values for the equation parameters (Table 4.3). Cells B2:B5 contain the loop information, where we can read that the simulation will run from day 0 to 10 (cell B2) with daily time steps (_stepsize or cell B3 set to 1).

In Fig. 4.13, cells E3:E6 calculate the solar irradiance based on Eq. 3.8 and cells E9:E12 the leaf photosynthesis based on Eq. 2.3. The double integration result to determine $A_{T,d}$ is stored in cell E16. Cells E21 and E22 implement Eq. 4.9a and b, respectively, to determine the total change in W_s and W_g within the current day. Cells E19 and E20 store the current W_s and W_g values, respectively. Both these values are used simultaneously by cells E21 and E22 to estimate the new weights of W_s and W_g for the next day.

	A	B	C
1	CONTROL		
2	max steps	10	
3	step size	1	
4	step		
5	criteria	=_step<=B2	
6			
7	PARAMETER:		
8	Am	1500	
9	k	0.5	
10	e	12	
11	ltd	10000000	
12	tsr	6	
13	tss	18	
14	LAI	2	
15	kg	1.98	
16	kd	0.1	
17	ks	0.05	
18	Yg	0.75	
19			
20	INITIAL		
21	W_s	20	
22	W_g	130	
23			

Fig. 4.12. Implementation of the simple crop growth model: setup of the loop section and the equation parameter values.

In Fig. 4.14, the prerun section (starting at cell G8 which is defined with cell name _prerun) contains the INI action which would initialize W_s and W_g (cells E19 and

109

E20, respectively) with the values from cells B21 and B22, respectively (Fig. 4.12).

	C	D	E	F
1		**MODEL**		
2		***Solar irradiance***		
3		t, hour		
4		n1	=Itd/(1800*(tss-tsr))	
5		n2	=SIN(PI()*(hour-tsr)/(tss-tsr))^2	
6		Io	=E4*E5	
7				
8		***Leaf photosyn.***		
9		L		
10		n3	=Am*e*k*Io*EXP(-k*L)	
11		n4	=Am+e*k*Io*EXP(-k*L)	
12		A_L	=E10/E11	
13				
14		***Daily canopy photosyn.***		
15		ITG result		
16		A_{Td}	=(12/44)*3600*LAI*E15/10^6	
17				
18		***Weights***		
19		W_s		
20		W_g		
21		ΔW_s	=(kd*Wg-kg*Ws)*_stepsize+ATd	
22		ΔW_g	=(kg*Ws*Yg-kd*Wg-ks*Wg)*_stepsize	
23				

Fig. 4.13. Implementation of the simple crop growth model: setup of the calculations section.

Determination of the daily canopy photosynthesis $A_{T,d}$ requires two integrations, so in the operation section (starting from cell G2, which is defined with cell name _operation), two successive ITG actions are specified: the first to integrate A_L (cell E12) with respect to L (from 0 to

LAI) and the second with respect to t (from sunrise to sunset) (Fig. 4.14).

	F	G	H	I	J	K	L	M	N
1		OPERATION							
2		ITG	=L	=AL	=E15	0	=LAI	FALSE	
3		ITG	=hour	=AL	=E15	=tsr	=tss		
4		UPD	=E19:E20	=E21:E22	1				
5									
6									
7		PRERUN							
8		INI	=E19:E20	=B21:B22					
9									
10									

Fig. 4.14. Implementation of the simple crop growth model: setup of the operation and prerun sections.

Table 4.4. Cell names defined in the worksheet to implement the simple crop growth model (BuildIt cell names are listed first).

Cell	Cell name	Cell	Cell name
B3	_stepsize	B14	LAI
B4	_step	B15	kg
B5	_criteria	B16	kd
G2	_operation	B17	ks
G8	_prerun	B18	Yg
P3	_read	E3	hour
P7	_write	E6	Io
B8	Am	E9	L
B9	k	E12	AL
B10	e	E16	ATd
B11	Itd	E19	Ws
B12	tsr	E20	Wg
B13	tss		

Note that cell M2 is set to FALSE and cell M3 is left blank (TRUE by default) so that the aforementioned double integration of A_L is performed correctly.

The last action to be performed in the operation section is the UPD action. This action updates the W_s and W_g weights based on Eq. 4.10. Also note that the UPD's n parameter (cell J4) is set to 1 so that the daily time step will only have one subinterval. Cell J4 should be set only to 1 (or left blank) because Eq. 4.9a shows that $A_{T,d}$ is obtained by integrating A_T from sunrise to sunset, representing one day.

A single UPD action is required to simultaneously update both the dry weights W_s and W_g. Like in the case of the prey-predator model, we cannot have two UPD actions to first update $W_{s,t}$ to $W_{s,t+1}$, then the second to update $W_{g,t}$ to $W_{g,t+1}$. This is because after the first UPD action, it is the future value of $W_{s,t+1}$ and not the current value of $W_{s,t}$ that is used to update $W_{g,t}$ to $W_{g,t+1}$ (Eq. 4.9b and 4.10b).

In Fig. 4.15, cell P3 is defined with the cell name _read so that the cell range P3:V3 instruct BuildIt to output the time step (day), daily changes to the storage and structural dry weights (ΔW_s and ΔW_g), current storage and structural dry weights (W_s and W_g), and the total dry weight ($W_s + W_g$). Cells R4:V4 are set to FALSE so that all weights are to be outputted before the operation section because these weights will change when the UPD action is executed.

Model results will be stored as a list starting from cell P7 because this cell is defined with the cell name _write.

Once model implementation is completed, click the "Start Simulation" from the BuildIt menu to start the simulation. The model output is as shown in Fig. 4.16.

O	P	Q	R	S	T	U	V	W
1	TO OUTPUT							
2	day	A_{Td}	ΔW_s	ΔW_g	W_s	W_g	W_s+W_g	
3	=_step	=ATd	=E21	=E22	=Ws	=Wg	=Ws+Wg	
4			FALSE	FALSE	FALSE	FALSE	FALSE	
5								
6	OUTPUT							
7								
8								
9								
10								

Fig. 4.15. Implementation of the simple crop growth model: setup of the what-to-output and output sections.

O	P	Q	R	S	T	U	V
1	TO OUTPUT						
2	day	A_{Td}	ΔW_s	ΔW_g	W_s	W_g	W_s+W_g
3	0	0	0	0	0	0	0
4			FALSE	FALSE	FALSE	FALSE	FALSE
5							
6	OUTPUT						
7	0	22.32653	-26.6	10.2	20	130	150
8	1	22.32653	5.208005	2.32389	15.72653	140.2	155.9265
9	2	22.32653	-4.87146	9.709194	20.93453	142.5239	163.4584
10	3	22.32653	5.744947	1.018703	16.06307	152.2331	168.2962
11	4	22.32653	-5.52818	9.397143	21.80802	153.2518	175.0598
12	5	22.32653	6.357328	-0.22177	16.27984	162.6489	178.9288
13	6	22.32653	-6.25236	9.252126	22.63717	162.4272	185.0643
14	7	22.32653	7.052524	-1.42045	16.38481	171.6793	188.0641
15	8	22.32653	-7.05352	9.26562	23.43734	170.2588	193.6962
16	9	22.32653	7.83901	-2.5987	16.38382	179.5245	195.9083
17	10	22.32653	-7.9421	9.432036	24.22283	176.9258	201.1486

Fig. 4.16. Simulation results from the simple crop growth model.

Notice that the daily canopy photosynthesis $A_{T,d}$ (column Q) is constant throughout the time steps. This is because the daily total solar irradiance $I_{t,d}$ is artificially kept constant at 10 MJ m^{-2} every day. Moreover, this crop growth model did not account for any leaf growth (and leaf death); thus, LAI would remain constant at 2 m^{-2} leaf m^{-2} ground.

Later, in the Part II of this book, we will learn how to build a more rigorous and comprehensive crop growth model; that is, one relying less on simplistic assumptions.

4.5   Exercises

1. As discussed in section 4.1, a time interval Δt in Eq. 4.1 that is too large risks a large estimation error in particular when the rate of change dx_t/dt varies widely within Δt. Consequently, it is recommended to divide a large time interval into two or more subintervals. The more number of subintervals, the lower the estimation error. But having too many subintervals can lead to excessive calculations which can considerably slow down model runs with little gains in estimation accuracy.

   So, how do we decide how many subintervals to have? Discuss by giving an example.

2. A plant has an initial weight of 50 g. Its weight increases at a rate of $(1 + t)$ g day^{-1}, where t is the number of days. The weight of the plant at time t can be determined by

   $$W_{t+\Delta t} = W_t + (1+t)\Delta t$$

where W_t and $W_{t+\Delta t}$ are the plant weights at time t and $t + \Delta t$, respectively; and Δt is the time interval.

Implement the loop and use the UPD and INI actions to determine the daily plant weight from $t = 0$ to 10 days at every $\Delta t = 1$ day interval.

Your model output should be a list of (t, W_t) values which are then plotted.

3. From Question 2, modify your model implementation so that your model uses the REP action, instead of the UPD action. Your model should still use the INI action (in conjunction with the REP action) and a loop for the iterative calculations to determine the daily plant weights.

   What are the benefits and disadvantages of using the UPD over the REP action?

4. Modify the simple plant growth model as shown in section 4.4, so that the leaf area index *LAI* is not a constant but is now determined by

   $$LAI = L_{AR} \times W_g$$

   where L_{AR} is the leaf area ratio (leaf area per unit structural dry weight; m^2 leaf g^{-1} C), and W_g is the structural dry weight (g C m^{-2} ground).

   Take L_{AR} as 0.02 m^2 leaf g^{-1} C, and the values for the rest of the model parameters are as before (Table 4.3).

   Determine the structural (W_g) and substrates (W_s) dry weights, *LAI*, and daily photosynthetic rate $A_{T,d}$ for $t = 0$ to 10 at daily intervals ($\Delta t = 1$).

   Repeat the above task but now for $t = 0$ to 10 at two-day intervals ($\Delta t = 2$).

What are the differences between the two sets of model outputs (the output when $\Delta t = 1$ with that when $\Delta t = 2$), and why?

Chapter 5. Scenarios, macros, and functions

5.1 Scenarios

In simulations, we are often interested to determine the predicted outcome of different scenarios. This is often done by changing the values of some of the model parameters and running the simulation again to obtain the new predicted outcome.

In section 4.3, the prey-predator model was used to simulate the population dynamics of both prey and predator. What if we double the growth rates for both the prey and predator? That is, what happens to the population dynamics when we change the growth rate for prey from 0.1 to 0.2 and for predator from 0.001 to 0.002?

Typically, we would manually change the value in cells B12 and B13 (Fig. 4.6a) to 0.2 and 0.002, respectively. We then run the model again and compare its output with the original outcome (with the lower growth rates). Or, alternatively, we can use BuildIt's functionality for running scenarios (Fig. 5.1).

Fig. 5.2 shows the changes to the implementation for the prey-predator model in Excel to include scenarios. The implementation is exactly the same as shown in Fig. 4.6c except for the addition of the scenario section, starting from cell H13. This cell is defined with the cell name `_scenario`.

Whenever we define a cell with the name `_scenario`, BuildIt will begin looking at pairs of cells. The search starts with the cell defined with `_scenario` and moves progressively right, examining each cell pairs. BuildIt stops searching only when it encounters the first blank cell.

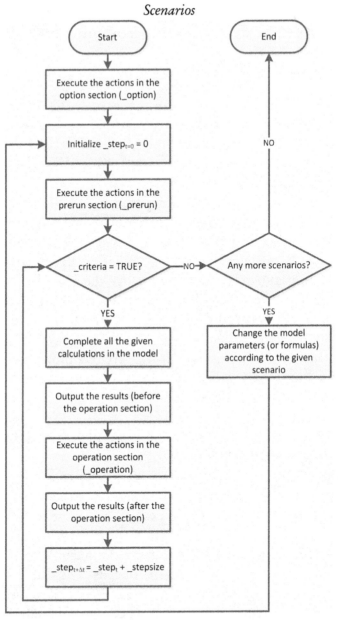

Fig. 5.1. Program flow of BuildIt which includes running scenarios.

G	H	I	J	K	L	M	N	O	P
1	OPERATION								
2	UPD	=E3:E4	=F3:F4	5					
3									
4									
5									
6									
7									
8	PRERUN								
9	INI	=E3:E4	=B9:B10						
10									
11									
12	SCENARIO								
13	=prey_growth	0.2	=pred_growth	0.002					
14	=prey_growth	0.2	=pred_growth	0.002	=prey_death	0.02	=pred_death	0.1	
15									

Fig. 5.2. Cell H13 is defined with the cell name _scenario to enable BuildIt to run scenarios in the prey-predator dynamics. Here, cell range H13:K13 define the second scenario and H14:O14 the third scenario.

As stated earlier, we defined _scenario to cell H13 in our example, so BuildIt will look at the pair of cells (H13, I13), followed by (J13, K13). BuildIt stops at (J13, K13) because the next cell L13 is blank. BuildIt treats each cell pair as (destination, source), where the contents from the source cell is copied to the destination cell.

To better explain this, let us now take the first cell pair (H13, I13) which has the following cell contents: (prey_growth, 0.2) (Fig. 5.2 and 5.3a). This instructs BuildIt to give prey_growth (cell name defined for cell B12; Table 4.2) a value of 0.2. The second cell pair (J13, K13), has the following contents: (pred_growth, 0.002) which indicates to BuildIt that pred_growth (defined for cell B13; Table 4.2) will have the value 0.002.

BuildIt first runs the simulation for the original growth rates of 0.1 and 0.001 for the prey and predator, respectively (Fig. 4.6a). Once the loop run stops, BuildIt checks for any given scenarios (Fig. 5.1). If there are no scenarios, the model run ends. But since we have specified at least one scenario (Fig. 5.2), BuildIt reads the specified scenario and doubles the growth rates for the prey and predator. BuildIt next initializes _step back to 0, executes the actions in the prerun section, then restarts the loop run to perform the iterative calculations. BuildIt repeats this process for the subsequent scenarios until all scenarios have been performed.

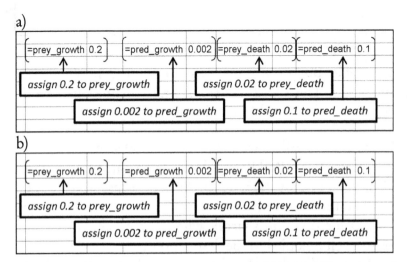

Fig. 5.3. In the scenario section, BuildIt reads pairs of (destination, source) cells so that each pair means the contents from the source cell is copied to the destination cell. Two additional scenarios are defined: a) the growth rates of the prey and predator are doubled, and b) both the growth and death rates for prey and predator are doubled.

In Fig. 5.2, we see that we have specified an additional two scenarios, so in total, we actually have three scenarios: the first is with the original growth rates for prey and predator. The second scenario, as stated earlier, is the doubling of their growth rates. And for the third scenario, we doubled the growth rates as well as the death rates for both the prey and predator.

In Fig. 5.2, the second scenario is specified in cell range H13:K13 and the third scenario in cell range H14:O14. For the third scenario, cell pairs (H14, I14) and (J14, K14) instruct BuildIt to double the growth rates of the prey and predator, respectively (like in the second scenario), and cell pairs (L14, M14) and (N14, O14) instruct BuildIt to double the death rates for the prey and predator, respectively (Fig. 5.3b).

Consequently, having three scenarios means we will have three sets of model output: one for the original scenario which is stored in cell range A7:C207, the second scenario in A210:C410, and the third scenario in A413:C613) (Fig. 5.4). The output of each scenario is separated by a blank row and a row header "Scenario no. {X}" where {X} is the scenario number; only the first scenario has no header.

In the model output (Fig. 5.5), we see that doubling the growth rates for both prey and predator changes significantly the dynamics of their population. Compared to the original growth rates, doubling the growth rate for prey did not cause a larger growth in its population. Instead, it showed a smaller increase in its population growth as the prey population was suppressed by the larger population growth for the predator.

However, the third scenario showed that doubling the prey growth would increase the prey population provided the death rate of the predator was increased, reducing the pressure on the prey population. We can also see the doubling the growth and death rates of the prey and predator caused greater fluctuations in the population dynamics.

	A	B	C	D
1	TO OUTPUT			
2	t	prey pop.	predator pop.	
3	201	20.6597	6.962815524	
4		FALSE	FALSE	
5				
6	OUTPUT			
7	0	50	15	
8	1	47.5509	14.98515659	
9	2	45.2352	14.93447949	
10	3	43.0603	14.85034588	
204	197	25.8372	12.48868611	
205	198	25.2312	12.18676837	
206	199	24.7142	11.88533948	
207	200	24.281	11.58573939	
208				
209	Scenario no. 2:			
210	0	50	15	
211	1	52.3898	15.79572908	
212	2	54.4374	16.70828053	
213	3	56.0282	17.73915327	
408	198	48.3313	12.3688799	
409	199	52.0104	12.9948082	
410	200	55.5972	13.75143366	
411				
412	Scenario no. 3:			
413	0	50	15	
414	1	45.207	14.94130274	
415	2	40.969	14.7470492	
416	3	37.3118	14.43877821	
611	198	18.0009	8.392119469	
612	199	18.6655	7.872577717	
613	200	19.5483	7.396445019	
614				

Fig. 5.4. Prey-predator model output for the three scenarios. Some rows of cells are hidden to show that the first scenario output listing is in cell range A7:C207, the second scenario in A210:C410, and the third scenario in A413:C613.

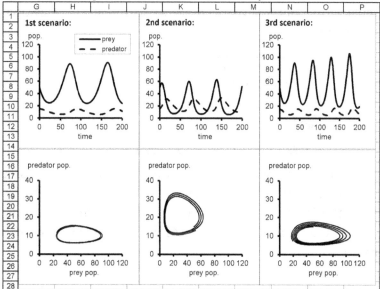

Fig. 5.5. Charts drawn from the model output from the prey-predator model to compare the results between the three scenarios.

Every time a scenario run ends, the contents of all (destination, source) cell pairs are restored to their original contents.

For instance, the growth rates of the prey and predator which we had doubled in the second scenario will be restored to their original values (as shown in Fig. 4.6a) when the second scenario run ends. Consequently, for the third scenario, we need to specify again to double the growth rates of the prey and predator.

When the third scenario run ends, the growth (and death) rates of the prey and predator will be restored to their original values.

BuildIt scenarios need not be limited to replacing a value (destination) with another value (source). You could

123

replace the contents of a cell with a formula such as shown in Fig. 5.6.

	A	B	C
40	SCENARIO		
41	=E5	=0.5*E3/E4	
42	=F6	=prey_growth	
43			

Fig. 5.6. BuildIt scenario copies the contents of the source cells to the respective destination cells, so it is possible to copy the formulas (not the results of the formulas) to the destination cells. Note: cell A41 is defined with the cell name _scenario.

In Fig. 5.6, it is the formula "=0.5*E3/E4" and not the result of the formula that is copied to cell E5. In other words, after the copy operation, cell E5 will have the formula "=0.5*E3/E4". Likewise, the formula: "=prey_growth" is copied to cell F6, not the value of prey_growth, so that cell F6 will always have the current value of prey_growth, not just its value at the time BuildIt performs the copy operation.

However, we must be careful to avoid copying formulas that would result in circular references after the copy operation, such as copying the formula "=0.5*E3/E4" to either cell E3 or E4.

5.2 Macros

Macros are computer programs that can contain a sequence of recorded key strokes, or they can contain computer code written in Visual Basic for Applications (VBA).

Macros extend the capabilities of Excel because users can write macros to automate certain repetitive or routine tasks. Macros can also be developed to perform complicated tasks that would otherwise be difficult to do in the two-dimensional layout of the spreadsheet table[2].

BuildIt can execute or run user-defined macros. Use the RUN action to instruct BuildIt to execute macros (Table 3.5). The RUN action has the following syntax:

```
RUN        {macro_name:s}
           {op:s=TRUE}
```

which executes the macro specified in macro_name, and the last parameter op is TRUE by default, so this action is always performed by BuildIt. Set the op parameter to FALSE to cancel the RUN execution.

The RUN action can be placed in the operation, prerun, or option section (Fig. 5.7).

Fig. 5.7 is the same as Fig. 5.1 except the former flowchart now includes the option section.

Any actions in the option section are executed only once by BuildIt, and their executions are done even before the loop counter, before _step initialization, and before the actions in the prerun section.

Similar to creating the operation or prerun section, to create the option section, simply define a cell with the name _option. If _option is defined, BuildIt will read the actions listed there and execute them in the order these actions are listed.

BuildIt supplies three utility macros, as shown in Table 5.1.

---

[2] Developing macros is beyond the scope of this book because it requires VBA programming proficiency. Many textbooks, however, are available in the market that teach VBA programming in Excel.

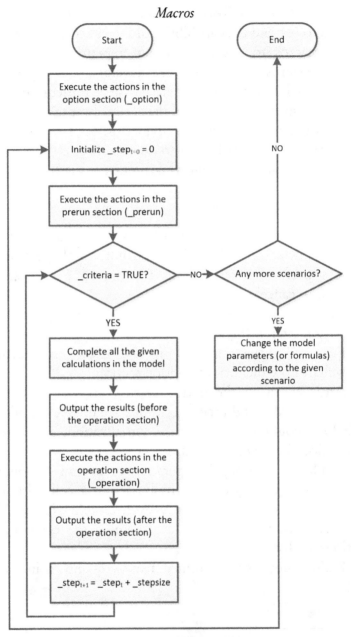

Fig. 5.7. The full program flow of BuildIt which includes the option section.

Table 5.1. BuildIt macros

Macro name	What the macro does
ClearOutput	Deletes the previous model output
DisableScreenUpdate	Switches off the automatic screen update (for speeding up model runs)
EnableScreenUpdate	Switches on the automatic screen update

The first macro, ClearOutput, is useful to clear any previous model output. Model simulations are often run many times, with each run producing its own output result. Consequently, in the Output section, you may have an overlap of model results, a mix of the old and latest model output. So, to ensure only the latest model output is stored or displayed, we might want to run the ClearOutput macro (Fig. 5.8).

	F	G	H	I	J
1		**OPTION**			
2		**RUN**	ClearOutput		
3					
4					

Fig. 5.8. BuildIt macro ClearOutput is specified in the option section, so that BuildIt deletes the previous model output at the start of a model run. Note: cell G2 is defined with the cell name _option.

Cell G2 is defined with the cell name _option, so this cell marks the start of the action list in the option section. In this case, only one action, RUN, is specified. The RUN action is used to instruct BuildIt to execute the

ClearOutput macro which will delete the previous model output at the start of a model run.

The other two macros, DisableScreenUpdate and EnableScreenUpdate, concern the computer screen refresh (Table 5.1). Turning off the screen update speeds up the simulation run because Excel will not refresh the cells values.

In other words, the computer screen appears "frozen". Though appearing to be stalled, the simulation is still running and cells are still being updated but the computer screen does not refresh to view their new updated values.

Turning on the screen update will then "unfreeze" the screen and all cells that are visible on the screen will have their correct and current values. These two macros are useful in particular for models which have intensive and lengthy calculations.

The DisableScreenUpdate macro is often used in the Option section so that the screen update is disabled due to lengthy calculations in the model.

Fig. 5.9 is one way this macro can be specified. This time, two macros are specified in the option section.

	F	G	H	I	J
1		**OPTION**			
2		**RUN**	ClearOutput		
3		**RUN**	DisableScreenUpdate		
4					

Fig. 5.9. BuildIt macros ClearOutput and DisableScreenUpdate are specified in the option section. Note: cell G2 is defined with the cell name _option.

The first macro, ClearOutput, deletes the previous model output, and the second macro,

`DisableScreenUpdate`, will suspend the automatic screen update to speed up model calculations.

You do not have to worry about forgetting to run the `EnableScreenUpdate` counterpart macro to re-enable the screen update after disabling it because BuildIt always ensures that the screen update is enabled before a simulation ends or even when a simulation run is aborted (such as by pressing the Esc or Control-C keys).

The macros `DisableScreenUpdate` and `EnableScreenUpdate` can be used as a pair to turn off the screen update before a lengthy calculation is performed, then turned back on the screen update after the lengthy calculation is completed such as shown in Fig. 5.10.

	F	G	H	I	J	K
10		**OPERATION**				
11		**RUN**	DisableScreenUpdate			
12		**UPD**	=A2	=A3	10	
13		**UPD**	=B2	=B3	10	
14		**RUN**	EnableScreenUpdate			
15						

Fig. 5.10. Macros `DisableScreenUpdate` and `EnableScreenUpdate` are used as a pair in the operation section to turn off the automatic screen updates to speed up lengthy calculations (by the two UPD actions), then to turn back on the screen updates, respectively. Note: cell G11 is defined with the cell name `_operation`.

Lastly, macro names are case-insensitive, so that `ClearOutput` can be specified in the action list as `clearoutput` or `Clearoutput`.

5.3 Functions

BuildIt supplies two custom functions: `interpolate` and `solve`.

The `interpolate` function performs linear interpolation between two values, and the `solve` function solves simultaneous linear equations. The `solve` function can also solve nonlinear equations, but it can only return a single solution at a time, even if there could be more than one solution to the nonlinear equations.

5.3.1 interpolate function

The `interpolate` function has the following specification:

```
interpolate(array_x, array_y, find_x)
```

where `array_x` and `array_y` are a list or array of (x, y) pairs. Given `find_x`, the interpolate function returns the corresponding y value.

If no exact match is found, this function performs a linear interpolation to obtain y.

It is important to remember that `array_x` must be sorted either ascendingly (from smallest to largest value) or descendingly (from largest to smallest value) before this function can be used reliably.

Consider a given (x, y) data set in Fig. 5.11a. The array of (x, y) values are given in cell range (A2:A10, B2:B10). The `interpolate` function is used in cell range E2:E5 to determine the y values for the corresponding x values given in cell range D2:D5. Also note that the array of x values in cell range A2:A10 have already been sorted, in this case, ascendingly.

a)

	A	B	C	D	E
1	x	y		x	est. y
2	0	10		2	=interpolate($A$2:$A$10,$B$2:$B$10,D2)
3	2	20		3	=interpolate($A$2:$A$10,$B$2:$B$10,D3)
4	4	50		15	=interpolate($A$2:$A$10,$B$2:$B$10,D4)
5	6	59		20	=interpolate($A$2:$A$10,$B$2:$B$10,D5)
6	8	62			
7	10	76			
8	12	89			
9	14	92			
10	16	99			
11					

b)

	A	B	C	D	E
1	x	y		x	est. y
2	0	10		2	20
3	2	20		3	35
4	4	50		15	95.5
5	6	59		20	#VALUE!
6	8	62			
7	10	76			
8	12	89			
9	14	92			
10	16	99			

Fig. 5.11. BuildIt's `interpolate` function used on (x, y) array in cell range (A2:A10, B2:B10): a) how the interpolate function is used in cell range E2:E5 to determine the corresponding y values for the x values in cell range D2:D5, and b) the results of the linear interpolation.

The first case (cell E2) returns the exact y value of 20 because an exact match was found (Fig. 5.11b).

The second case (cell E3) fails to find an exact match, so this function linear interpolates between (2, 20) and (4, 50) to give $y = 35$ for $x = 3$.

Similarly, in the third case (cell E4), linear interpolation was performed between (14, 92) and (16, 99) to given $y = 95.5$ for $x = 15$.

The last case (cell E5), however, returns an error value because $x = 20$ lies beyond the range of x array. If `find_x` is outside the `array_x` range, the error value "#VALUE!" is returned.

5.3.2   `solve` function

The `solve` function has the following specification:

```
solve(lhs, rhs, variables)
```

where `lhs` and `rhs` are the left and right hand sides of the equations, and `variables` are the equation variables you want to solve.

The `solve` function uses the Newton iteration method to find roots of nonlinear simultaneous equations, and the best way to understand how to use the `solve` function is by considering the following example.

We wish to determine the values for the variables w, x, y, and z such that:

$$w^3 + 2w^2 + 3w + 4 = 12.828$$
$$wx + xy + yz = -3.919$$
$$w^2 + 2wx + x^2 = 1$$
$$w + x + y - z = -3.663$$

Fig. 5.12 shows one way the `solve` function is used to estimate the values for the variables w, x, y, and z.

Best-guessed initial values for *w*, *x*, *y*, and *z* are provided in cells B2, B3, B4, and B5, respectively. The left hand side of the equations (lhs) are entered in cells B8:B11 and on the right hand side of the equations (rhs), the corresponding constants are entered in cells C8:C11.

a)

	A	B	C
1		**Initial values**	**Estimated values**
2	w	1	=solve(B8:B11,C8:C11,B2:B5)
3	x	1	=solve(B8:B11,C8:C11,B2:B5)
4	y	1	=solve(B8:B11,C8:C11,B2:B5)
5	z	-1	=solve(B8:B11,C8:C11,B2:B5)
6			
7	**No.**	**Eqn.**	**Constants**
8	1	=B2^3+2*B2^2+3*B2+4	12.828
9	2	=B2*B3+B3*B4+B4*B5	-3.919
10	3	=B2^2+2*B2*B3+B3^2	1
11	4	=B2+B3+B4-B5	-3.663

b)

	A	B	C
1		**Initial values**	**Estimated values**
2	w	1	1.249990148
3	x	1	-0.249990148
4	y	1	-3.329959069
5	z	-1	1.333040931
6			
7	**No.**	**Eqn.**	**Constants**
8	1	10	12.828
9	2	1	-3.919
10	3	4	1
11	4	4	-3.663

Fig. 5.12. BuildIt's solve function: a) setting up the worksheet to estimate the values for the variables *w*, *x*, *y*, and *z*, and b) the returned results in cells C2:C5.

To enter the `solve` formula, you need to first select cells C2:C5 and enter:

= solve(B8:B11,C8:C11,B2:B5)

where the cells B8:B11 are the left hand side of the equations, cells C8:C11 the right hand side, and cells B2:B5 the variables w, x, y, and z that we wish to determine their values.

Once the formula is typed, hit the Ctrl + Shift + Enter keys (not just the Enter key). This is because `solve` is an array function that can simultaneously return more than one value. Since we are dealing with four variables in this example, the `solve` function will return four values.

Once you hit the Ctrl + Shift + Enter keys, Excel inserts the curly braces in the formula you had earlier typed to give:

{ = solve(B8:B11,C8:C11,B2:B5)}

in cells C2:C5 (Fig. 5.13).

The results are as shown earlier in Fig. 5.12b. It is important to give good estimates to the variables' initial values. Initial values that are too far from their true values can lead to unacceptably inaccurate estimates.

A set of equations that includes one or more nonlinear equations can have more than one solution. The `solve` function unfortunately does not simultaneously return all solutions. `solve` only finds the solution closest to the given initial estimates. For instance, consider the following set of two equations:

$$x^2 - 2y = 5$$
$$x - y = 1$$

that has two solutions to (x, y): (-1, -2) and (3, 2). Note that the first equation in the set is a nonlinear equation.

If we give the initial estimates to (x, y) as $(-1, -1)$, `solve` returns the solution $(-1, -2)$, but if their initial estimates were instead $(2, 2)$, `solve` returns $(3, 2)$.

Consequently, `solve` is more reliable to solve linear equations.

	C2	▾	f_x {=solve(B8:B11,C8:C11,B2:B5)}
◢	A	B	C
1		Initial values	Estimated values
2	w	1	=solve(B8:B11,C8:C11,B2:B5)
3	x	1	=solve(B8:B11,C8:C11,B2:B5)
4	y	1	=solve(B8:B11,C8:C11,B2:B5)
5	z	-1	=solve(B8:B11,C8:C11,B2:B5)
6			
7	No.	Eqn.	Constants
8	1	=B2^3+2*B2^2+3*B2+4	12.828
9	2	=B2*B3+B3*B4+B4*B5	-3.919
10	3	=B2^2+2*B2*B3+B3^2	1
11	4	=B2+B3+B4-B5	-3.663

Fig. 5.13. The `solve` function is entered as an array formula to simultaneously return four values in cells C2:C5. Note the automatic insertion of the curly braces (as indicated by the arrows) by Excel in the array formula when we hit the Ctrl+Shift+Enter keys.

5.3.3   A note on #NAME? and link errors

BuildIt's functions, `interpolate` and `solve`, work as long as BuildIt is loaded in Excel. If the BuildIt add-in file is deleted from your computer or moved to another location, these two functions will fail and return the #NAME? error value. Likewise, if you unload BuildIt from Excel (*see* Appendix A), these functions will also fail to work.

More likely is this problem occurs when you open a workbook containing these BuildIt functions on different computers. These functions may work as expected on one computer but yet fail to work when the same workbook is opened on another computer, even if the second computer has BuildIt installed.

Excel will give an error message that one or more external links in the workbook cannot be updated. This error happens because BuildIt was not installed in the same location in your second computer compared to the first computer.

To fix this external link error, choose the "Fix #NAME? Error" command from BuildIt's menu (Fig. A.3 and A.5). This command will repair the "broken" BuildIt links in the workbook.

Note that you have to use the "Fix #NAME? Error" command each time you open the workbook on different computers. This is an issue with all user-defined functions, not just BuildIt's.

5.4   Summary of using BuildIt

Before we proceed to the next part of this book, let us summarize the steps in using BuildIt. Recall that BuildIt overcomes some of Excel's weaknesses mainly by providing: 1) a loop for iterative calculations and 2) actions to perform certain specific tasks not possible or difficult to do in native Excel.

A loop needs to be setup if a model requires the same set of calculations to be repeatedly performed. BuildIt requires three key information about the loop, which are:

    a) Loop counter. Define a cell with the name _step. BuildIt will use this cell exclusively to maintain the loop. BuildIt will initialize _step with 0, then increment it at the end of every loop cycle.

b) Interval size. Define a cell with the name _stepsize, and specify the size of the interval. At the end of every loop cycle, BuildIt will increment _step by _stepsize.

c) Loop criteria. Define a cell with the name _criteria, and specify the logical condition that would end the loop run. As long as _criteria is TRUE, the loop will run – a FALSE value will terminate the loop run.

BuildIt supplies 12 actions, and each of these actions performs a specific task often needed in modeling (Table 3.5). You use these actions by enumerating them in the prerun and operation sections.

The prerun section begins at the cell defined with the name _prerun and the operation section with _operation. Any actions listed in the prerun section are executed by BuildIt only once, at the start of a model run. But actions listed in the operation section are executed once per loop cycle (Fig. 5.7).

The what-to-output and output sections instruct BuildIt what you want listed in the model output and where this model output should be stored or displayed in the worksheet. The what-to-output section begins with the cell defined with the name _read. It lists the variables or parameters you wish to output.

Model output occurs twice: once before BuildIt executes the actions in the operation section and once after the operation section (Fig. 5.7). To instruct BuildIt to output selected variables before the operation section, specify FALSE for these variables, or else to TRUE if their output is to occur after the operation section (*see* Fig. 4.3).

Variables where their values are altered within the operation section should typically be outputted before the operation section.

Model output appears as a list and starts at the output section, defined with the cell name _write.

BuildIt supports model scenarios (section 5.1). Having completed the first model run (which is considered as the first scenario), BuildIt checks if the _scenario cell name has been defined for any cell (Fig. 5.7). If it has, BuildIt reads the one or more pairs of (destination, source) cells to copy the contents from the source cells to the corresponding destination cells, then runs the model again using these modified cell contents. BuildIt will continue to run successive scenarios until all scenarios have been completed.

More precisely, actions in the prerun section will be executed at the start of each model scenario, but those in the option section will only be executed once, at the start of a model run.

The option section is defined by the cell name _option. This section, like the prerun and operation sections, contain the list of actions BuildIt should execute but only once when the model is first run, as stated earlier. Even if there are any model scenarios, the actions in the option section will not be executed again. Consequently, the RUN action is typically listed in this section to run any macros to perform "one-off" tasks (such as the ClearOutput macro to delete any old model output) before calculations begin in the model.

You will typically adhere the following general steps in building your models in Excel:
1. Implement all model calculations,
2. Set up a loop for the repetitive calculations,

3. Set up additional tasks needed such as numerical integrations, initialization of variables, and solving differential equations, as well as copying and manipulation of cell ranges, and

4. Set up the model output; that is, to specify what to output from the model and where the model output should appear in the worksheet.

5. Run the model.

5.5 Exercises

1. Modify the prey-predator model so that the model output from the previous model run is always deleted when the model first runs.

2. Modify the simple plant growth model from section 4.4 to include the following three scenarios:

   a) Scenario 2: Initial W_s and W_g = 10 and 65 g C m^{-2} ground, respectively.

   b) Scenario 3: Leaf area index LAI is not a constant but determined by

   $$LAI = L_{AR} \times W_g$$

   where L_{AR} is the leaf area ratio (leaf area per unit structural dry weight; m^2 leaf g^{-1} C), and W_g is the structural dry weight (g C m^{-2} ground). In this scenario, L_{AR} = 0.02 m^2 leaf g^{-1} C.

   c) Scenario 4: Both scenario 2 and 3; that is, initial W_s and W_g = 10 and 65 g C m^{-2} ground, respectively, and $LAI = L_{AR} \times W_g$ with L_{AR} = 0.02 m^2 leaf g^{-1} C.

Recall that the first model run is considered as the first scenario.

3. Given the following (x, y) pairs of values:

$$(2,5),\ (6,17),\ (9,29),\ (11,35),\ (15,14),\ (17,0)$$

use the `interpolate` function to determine y for $x = 4$, 10, and 16.

The function `interpolate` returns the error value #VALUE! if we try to determine y for $x = 0$ because 0 lies beyond the range of x values. Implement one extrapolation method in Excel that you would use to estimate y for $x = 0$.

4. Use the `solve` function to solve the following equations:

a) Determine x and y such that

$$2x + y = 15$$
$$-4x + 6y = -2$$

b) Determine x and y such that

$$x + y = 1$$
$$x^2 - 2x - y = 1$$

c) Determine x, y, and z such that

$$x + y + 2z = 10$$
$$-3x - 2y + 10z = 2$$
$$x + 4y - 3z = 12$$

PART II
Building a crop growth model

Chapter 6. Meteorology component

6.1 Overview of the crop growth model

The model we will build in Excel is a generic crop growth model, so-called because the crop is a hypothetical crop, having properties broadly or generally true for annual C3 crops.

Recall that the main purpose of this book is not to discuss the science of crop growth or to describe how the various processes in crop growth can be described or translated mathematically into a model.

Instead, the following chapters will show how Excel, with the help from BuildIt, can be used to build mathematical models such as a crop growth model. But at the same time, the crop growth model we will build should not be trivial or simplistic so that certain key implementation steps required in model building can be highlighted.

At the end, this book's crop growth model can serve as a template or foundation upon which we build a more complex crop growth model or a growth model intended for a specific crop.

The generic crop growth model we will build is simply called as gcg (generic crop growth) model. gcg has five main components: 1) meteorology, 2) photosynthesis, 3) energy balance, 4) soil water, and 5) growth development (Fig. 6.1).

These five components do not include other factors that are also important to crop growth such as nutrient requirement, damages from pests, diseases, and weeds, and management practices. To avoid the distraction of having to build such a complex crop growth model, we will concentrate on building a crop growth model where crop growth is only affected by meteorological factors and availability of water. Other factors are assumed to be non-limiting to crop growth.

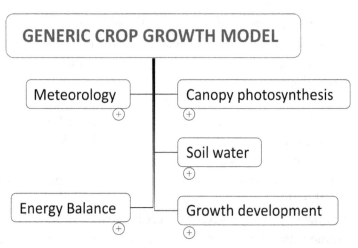

Fig. 6.1. Overview of the gcg (generic crop growth) model, with its five components: meteorology, photosynthesis, energy balance, soil water, and growth development. Other factors important to crop growth (such as nutrients, management, and pests and diseases) are assumed to be non-limiting.

6.2 Equations

This section documents all the equations used in the meteorology component of the gcg model. The equations are taken (and some adapted) from Kaplanis (2005), Ortega-Farias *et al.* (2000), Ephrath *et al.* (1996), Goudriaan and van Laar (1994), de Jong (1980), and Frere and Popov (1979).

The various properties calculated by the meteorology component is visually depicted in Fig. 6.2.

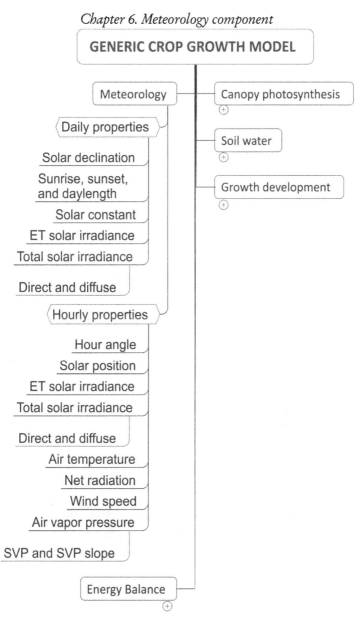

Fig. 6.2. The meteorology model component. ET and SVP denote extraterrestrial and saturated air vapor pressure, respectively.

6.2.1 Daily properties

Daily properties refer to meteorological properties for a given day of year (or date).

6.2.1.1 Solar declination

$$\delta = -0.4093 \cos\left[\frac{2\pi(t_d + 10)}{365}\right] \tag{6.1}$$

where δ is the solar declination (radians); and t_d is the day of year.

6.2.1.2 Sunrise, sunset and daylength

$$t_{sr} = 12 - \frac{12}{\pi}\mathrm{acos}(-a/b) = 24 - t_{ss} \tag{6.2a}$$

$$a = \sin\lambda\sin\delta \tag{6.2b}$$

$$b = \cos\lambda\cos\delta \tag{6.2c}$$

$$t_{ss} = 12 + \frac{12}{\pi}\mathrm{acos}(-a/b) \tag{6.3}$$

$$DL = \frac{24}{\pi}\mathrm{acos}(-a/b) = t_{ss} - t_{sr} \tag{6.4}$$

where t_{sr}, t_{ss} and DL are the sunrise, sunset, and daylength, respectively (all expressed in hours); δ is the solar declination (radians); and λ is the site latitude (radians).

6.2.1.3 Solar constant

$$I_c = 1370\left\{1 + 0.033\cos\left[\frac{2\pi(t_d - 10)}{365}\right]\right\} \tag{6.5}$$

where I_c is the solar constant (W m^{-2}); and t_d is the day of year.

6.2.1.4 Daily extraterrestrial solar irradiance

$$I_{et,d} = 3600 \cdot I_c \cdot \frac{24}{\pi} \left[a \cdot \text{acos}\left(-\frac{a}{b}\right) + b\sqrt{1 - \left(\frac{a}{b}\right)^2} \right] \quad (6.6)$$

where $I_{et,d}$ is the daily extraterrestrial solar irradiance (J m^{-2} day^{-1}); I_c is the solar constant (W m^{-2}); and parameters a and b are from Eq. 6.2b and c, respectively.

6.2.1.5 Daily total solar irradiance

$$I_{t,d} = I_{et,d}\left(b_0 + b_1 \frac{s}{DL}\right) \quad (6.7)$$

where $I_{t,d}$ is the daily total solar irradiance (J m^{-2} day^{-1}); $I_{et,d}$ is the extraterrestrial solar irradiance (J m^{-2} day^{-1}); b_0 and b_1 are empirical Angstrom coefficients, where their values depend on site location; s is the duration of sunshine hours (hours); and DL is the daylength (hours).

Note that the approximate values for the two Angstrom coefficients can also be obtained from Table 6.1.

Direct and diffuse components

$$\frac{I_{df,d}}{I_{t,d}} = \begin{cases} 1 & A < 0.07 \\ 1 - 2.3(A - 0.07)^2 & 0.07 \le A < 0.35 \\ 1.33 - 1.46A & 0.35 \le A < 0.75 \\ 0.23 & 0.75 \le A \end{cases} \quad (6.8)$$

$$I_{dr,d} = I_{t,d} - I_{df,d} \quad (6.9)$$

where $I_{df,d}$ is the daily diffuse solar irradiance (J m^{-2} day^{-1}); $I_{dr,d}$ is the daily direct solar irradiance (J m^{-2} day^{-1}); $I_{t,d}$ is the daily total solar irradiance (J m^{-2} day^{-1}); $I_{et,d}$ is the extraterrestrial solar irradiance (J m^{-2} day^{-1}); and $A = I_{t,d}/I_{et,d}$.

Table 6.1. The Angstrom coefficients b_0 and b_1 used for calculating daily solar radiation for different climate zones.

Climate zones	b_0	b_1
Cold or temperate	0.18	0.55
Dry tropical	0.25	0.45
Humid tropical	0.29	0.42

6.2.2 Hourly properties

Hourly properties refer to meteorological properties for a given hour and day of year (or date and time).

6.2.2.1 Hour angle

$$\tau = \frac{\pi}{12}(t_h - 12) \tag{6.10}$$

where τ is the hour angle (radians); and t_h is the local solar time (hours).

Note: hour angles will be negative for before solar noon ($t_h < 12$), positive for after solar noon ($t_h > 12$), and zero at exactly solar noon.

6.2.2.2 Solar position

Solar inclination and solar height

$$\theta = \text{acos}(a + b \cdot \cos \tau) \tag{6.11}$$

where θ is the solar inclination (angle from vertical) (radians); τ is the hour angle (hours); and parameters a and b are from Eq. 6.2b and c, respectively.

Note: rather than its angle from vertical, solar angle can be specified alternatively as β, the solar height or solar elevation, which is the solar angle from horizontal, determined by $\pi/2 - \theta$ (radians).

Solar azimuth

$$\phi = \begin{cases} \pi - \text{acos}\left(\dfrac{\sin \lambda \cdot \sin \beta - \sin \delta}{\cos \lambda \cdot \cos \beta} \right) & t_h \leq 12 \\[3mm] \pi + \text{acos}\left(\dfrac{\sin \lambda \cdot \sin \beta - \sin \delta}{\cos \lambda \cdot \cos \beta} \right) & t_h > 12 \end{cases} \tag{6.12}$$

where ϕ is the solar azimuth (angle from North in a clockwise direction) (radians); δ is the solar declination (radians); τ is the hour angle (hours); λ is the site latitude (radians); and β is the solar height or solar elevation (radians).

6.2.2.3 Extraterrestrial solar irradiance

$$I_{et} = I_c \sin \beta = I_c \cos \theta \tag{6.13}$$

where I_{et} is the hourly extraterrestrial solar irradiance (W m^{-2}); I_c is the solar constant (W m^{-2}); θ is the solar inclination (radians); and β is the solar height (or solar elevation) (radians).

Note: $\beta = \pi/2 - \theta$.

6.2.2.4 Total solar irradiance

$$I_t = a \cdot \psi - (b \cdot \psi) \cos\left(\pi \frac{t_h}{12}\right) \tag{6.14a}$$

$$\psi = \frac{\pi I_{t,d}/86400}{a \cdot a\cos(-a/b) + b\sqrt{1 - (a/b)^2}} \tag{6.14b}$$

where I_t is the hourly total solar irradiance (W m^{-2}); $I_{t,d}$ is the daily total solar irradiance (J m^{-2} day^{-1}); t_h is the local solar hour (hours); and δ is the solar declination (radians); and parameters a and b are from Eq. 6.2b and c, respectively.

Direct and diffuse components

$$\frac{I_{df}}{I_t} = \begin{cases} 1 & B \leq 0.22 \\ 1 - 6.4\left(\dfrac{I_t}{I_{et} - 0.22}\right)^2 & 0.22 < B \leq 0.35 \\ 1.47 - 1.66B & 0.35 < B \leq K \\ R & K < B \end{cases} \tag{6.15a}$$

$$R = 0.847 - 1.61\sin\beta + 1.04\sin^2\beta \tag{6.15b}$$

$$K = \frac{1.47 - R}{1.66} \tag{6.15c}$$

$$I_{dr} = I_t - I_{df} \tag{6.16}$$

where I_{df} is the hourly diffuse solar irradiance (W m^{-2}); I_{dr} is the hourly direct solar irradiance (W m^{-2}); I_t is the hourly total solar irradiance (W m^{-2}); I_{et} is the hourly extraterrestrial solar irradiance (W m^{-2}); β is the solar height (or solar elevation) (radians); and $B = I_t/I_{et}$.

Note: $\beta = \pi/2 - \theta$, where θ is the solar inclination.

6.2.2.5 Air temperature

Air temperature is calculated using one of three equations, depending on the current local solar time t_h (hours). If $t_h < (t_{sr} + 1.5)$, then

$$T_a = T_{set} + \frac{(T_{min} - T_{set})(t_h + t_{sr})}{(t_{sr} + 1.5) + t_{sr}} \qquad (6.17a)$$

or if $(t_{sr} + 1.5) \leq t_h \leq t_{ss}$, then

$$T_a = T_{min} + (T_{max} - T_{min}) \sin\left[\frac{\pi(t_h - t_{sr} - 1.5)}{DL}\right] \qquad (6.17b)$$

else if $t_h > t_{ss}$ then

$$T_a = T_{set} + \frac{(T_{min} - T_{set})(t_h - t_{ss})}{(t_{sr} + 1.5) + t_{sr}} \qquad (6.17c)$$

with

$$T_{set} = T_{min} + (T_{max} - T_{min}) \sin\left[\frac{\pi(t_{ss} - t_{sr} - 1.5)}{DL}\right] \qquad (6.17d)$$

where T_a is the air temperature (°C); T_{min} and T_{max} are the minimum and maximum air temperature (°C), respectively; t_{sr} and t_{ss} are the times of sunrise and sunset, respectively (hours); DL is the daylength (hours); and T_{set} is the air temperature (°C) at sunset (t_{ss}).

Note: minimum air temperature is assumed to occur 1.5 hours after sunrise.

6.2.2.6 Net radiation

$$R_n = (1 - p)I_t + R_{nL} \qquad (6.18a)$$

151

$$R_{nL} = 0.98\sigma T_{aK}^4 \left[1.31\left(e_a/T_{aK}\right)^{1/7} - 1\right] \tag{6.18b}$$

where R_n is the net radiation (W m^{-2}); R_{nL} is the net longwave radiation; I_t is the hourly total solar irradiance (W m^{-2}); p is the surface albedo (typically, 0.15); e_a is the air vapor pressure (mbar); T_{aK} is the air temperature (K); and σ is the Stefan-Boltzmann constant (5.67 x 10^{-8} W m^{-2} K^{-4}).

Note: T_{aK} (in K) = T_a (in °C) + 273.15.

6.2.2.7 Saturated air vapor pressure

$$e_s\llbracket T_a \rrbracket = 6.1078 \exp\left[17.269T_a/\left(T_a + 237.3\right)\right] \tag{6.19}$$

where $e_s\llbracket T_a \rrbracket$ is the saturated vapor pressure (mbar) at air temperature T_a (°C).

6.2.2.8 Slope of the saturated vapor pressure curve

$$\Delta = \frac{25029.4 \exp\left[17.269T_a/\left(T_a + 237.3\right)\right]}{\left(T_a + 237.3\right)^2} \tag{6.20}$$

where Δ is the slope of the saturated vapor pressure curve (mbar K^{-1} or equivalently, mbar °C^{-1}) at air temperature T_a (°C).

6.2.2.9 Air vapor pressure

Air vapor pressure is calculated similar to that for the saturated air vapor pressure:

$$e_a = 6.1078 \exp\left[\frac{17.269T_{d(cal)}}{T_{d(cal)} + 237.3}\right] \tag{6.21a}$$

where e_a is the air vapor pressure (mbar); $T_{d(cal)}$ is the calculated dew point temperature (°C), determined by:

$$T_{d(cal)} = \begin{cases} T_a & T_a < T_{d(\max)} \\ T_{d(\max)} & T_a \geq T_{d(\max)} \end{cases} \tag{6.21b}$$

where $T_{d(max)}$ is maximum dew point temperature (°C).

Air vapor pressure deficit

Air vapor pressure deficit is simply the difference between saturated and current air vapor pressure; that is,

$$D = e_s \llbracket T_a \rrbracket - e_a \tag{6.22}$$

where D is the vapor pressure deficit (mbar); e_a is the air vapor pressure (mbar); and $e_s \llbracket T_a \rrbracket$ is the saturated air vapor pressure (mbar) at air temperature T_a (°C).

6.2.2.10 Relative humidity

$$RH = 100 \frac{e_a}{e_s \llbracket T_a \rrbracket} \tag{6.23}$$

where RH is the relative humidity (%); $e_s \llbracket T_a \rrbracket$ is the saturated vapor pressure (mbar) at air temperature T_a (°C); and e_a is the air vapor pressure (mbar).

6.2.2.11 Wind speed

It assumed that the wind speed varies sinuously within a day as

$$u = \begin{cases} u_{\min} & t_h < (t_{sr} + 1.5) \text{ or } t_h > (t_{ss} + 1.5) \\ u_{\min} + u_\Delta & (t_{sr} + 1.5) \leq t_h \leq (t_{ss} + 1.5) \end{cases} \tag{6.24a}$$

$$u_\Delta = (u_{max} - u_{min}) \sin\left[\pi(t_h - t_{sr} - 1.5)/DL\right] \qquad (6.24b)$$

$$u_{min} = 0.5u_d \qquad (6.24c)$$

$$u_{max} = u_{min} - \left[12\pi(u_{min} - u_d)/DL\right] \qquad (6.24d)$$

where u is the wind speed (m s^{-1}) for the given hour (t_h); u_d is the daily total wind run (typically expressed as km day^{-1} but must be converted to m s^{-1}); u_{min} and u_{max} are the minimum and maximum wind speed for the day, respectively (m s^{-1}); t_h is the local solar hour (hour); t_{sr} and t_{ss} are the hour of sunrise and sunset, respectively (hour); and DL is the daylength (hour).

Note: It is assumed that minimum wind speed is half of u_d. Eq. 6.24 is formulated in such a way so that integrating Eq. 6.24 over 24 hours and dividing the integration result by 24 would give the average wind speed for the whole day, or u_d.

6.3 Measured parameters

The following are parameters that must be measured or supplied to the meteorology model component:

1. Site latitude (λ) and day of year (t_d).
2. Daily weather properties for the given site: sunshine hours (s), the minimum and maximum air temperatures (T_{min} and T_{max}), and daily total wind run (u_d). Although not required at the moment, data on rainfall amount is also needed when we later determine the soil water content.

   Daily weather properties are typically obtained from the weather station at the site.

3. The maximum dew point temperature $Td_{(max)}$ at the site. This value is typically obtained by examining historical weather data to determine the persisting maximum dew point temperature at the given site. This value would also vary according to seasons. Nonetheless, in humid, tropical countries like Malaysia, the maximum dew point temperature is rather constant throughout the year.

6.4 Implementation in Excel

6.4.1 Workbook and worksheets

The gcg model will be implemented in a single workbook, and this workbook will contain several worksheets, where each worksheet either stores tabulated data or implements individual components of the model. (Fig. 6.1).

At the moment, four worksheets are required. Create these four worksheets, and give them the following names: *Serdang, Meteorology, Control,* and *Output* (Fig. 6.3).

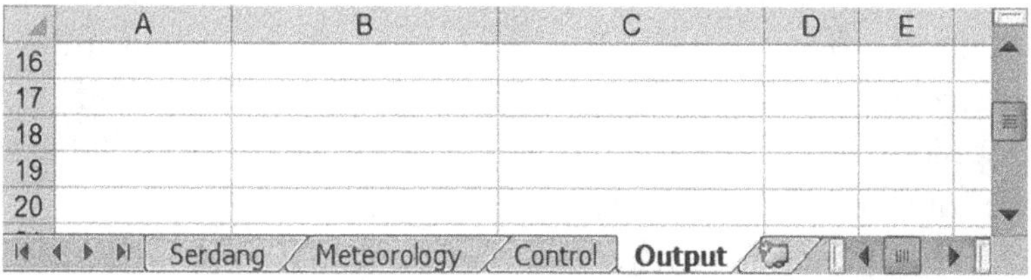

Fig. 6.3. Create four worksheets (and name them Serdang, Meteorology, Control, and Output, respectively) in the gcg model workbook.

The Serdang worksheet contains the daily weather data for Serdang town in Malaysia, the Meteorology worksheet contains the various equations as listed earlier in this chapter, and the last two worksheets, Control and Output, contain information about the loop and the output listing, respectively. Note that these two last worksheets are temporary and will change when we build the other model components.

6.4.2   Weather data

The gcg model requires only the daily weather properties. Hourly weather properties will be generated from daily weather properties.

Looking up a value from the weather table is more efficient and convenient if all the site's weather data are stored in a single worksheet. This includes storing all the site's weather data for one or more years.

For example, the entire weather data for Serdang, Malaysia for the years 1985 to 2004 should be stored in a single worksheet.

Do not break up the weather data set, for instance, by storing each year's data in a separate worksheet or workbook.

In this book, the sample weather data from Serdang, Malaysia (from 1985-2004) will be used as an example. Instructions of where to download BuildIt and its supplementary files (including the Serdang weather file) can be found in Appendix A.

The Serdang weather file "serdang.xls" obtained from the website is formatted such that the first three columns (A, B, and C) are for year, month and day, respectively (Fig. 6.4). Note that the months in column B are represented by values where 1 = January, 2 = February, and so on until 12 = December.

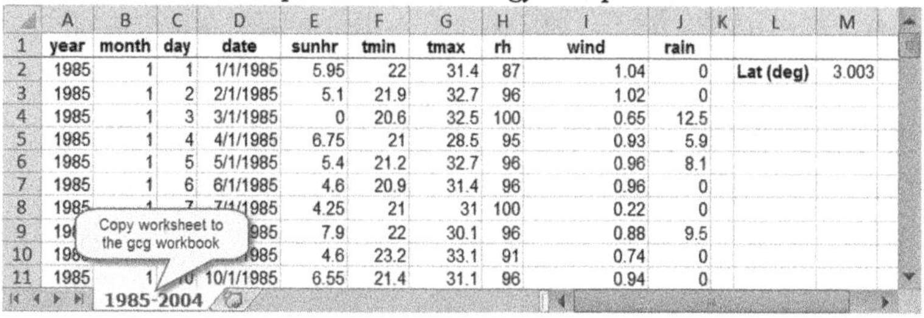

	A	B	C	D	E	F	G	H	I	J	K	L	M
1	year	month	day	date	sunhr	tmin	tmax	rh	wind	rain			
2	1985	1	1	1/1/1985	5.95	22	31.4	87	1.04	0		Lat (deg)	3.003
3	1985	1	2	2/1/1985	5.1	21.9	32.7	96	1.02	0			
4	1985	1	3	3/1/1985	0	20.6	32.5	100	0.65	12.5			
5	1985	1	4	4/1/1985	6.75	21	28.5	95	0.93	5.9			
6	1985	1	5	5/1/1985	5.4	21.2	32.7	96	0.96	8.1			
7	1985	1	6	6/1/1985	4.6	20.9	31.4	96	0.96	0			
8	1985	1	7	7/1/1985	4.25	21	31	100	0.22	0			
9	19	Copy worksheet to the gcg workbook		985	7.9	22	30.1	96	0.88	9.5			
10	19			985	4.6	23.2	33.1	91	0.74	0			
11	1985	1		10/1/1985	6.55	21.4	31.1	96	0.94	0			

1985-2004

Fig. 6.4. Download the file "serdang.xls" which contains the weather data of Serdang from 1984 to 2004. The weather data are kept in the worksheet named *1984-2004*.

The fourth column (column D) combines the year, month, and day into a date representation. Different countries specify dates in differing formats. For instance, a date written as "02/12/08" may mean 2 December 2008, 12 February 2008, or even 12 August 2002, depending on the computer's regional settings. To avoid this confusion, especially when the model is run on different computer systems, Excel's DATE function will be used to correctly specify the intended date. The DATE(year,month,day) function has three arguments which correspond to the year, month and day, respectively.

Cell D2, for instance, contains the formula: "=DATE(A2, B2, C2)" which returns "01-01-85" or equivalently, Jan. 1, 1985. Likewise, cell D3 contains the formula "=DATE(A3, B3, C3)" to return "02-01-85" to denote Jan. 2, 1985. In another computer with different regional settings, cell D3 may instead display "01-02-85" but because the DATE function is used in cell D3, this cell will still denote the intended date: Jan. 2, 1985.

This date column (column D) is essential because it is the key or unique identifier used to lookup the weather or meteorological properties based on a given date.

Columns E, F, G, H, and J contain information about the sunshine hours (hours), minimum and maximum air temperature (°C), relative humidity (%), total daily wind run or wind speed (m s^{-1}), and amount of rain (mm), respectively. Lastly, the value in cell M2 is the latitude (in degrees) of the Serdang site.

Note that the text in cell L2 and in cells A1:J1 are merely text labels for documentation and to help in model readability.

Once you have downloaded the Serdang weather file ("serdang.xls"), you need to transfer its weather data into the gcg workbook. To do this, you can either: 1) make a copy of the 1984-2004 worksheet (in "serdang.xls") and transfer the copy to the gcg workbook or 2) select all cells in the 1984-2004 worksheet (in "serdang.xls") and paste them in the blank Serdang worksheet in the gcg workbook. Whichever method you use, ensure the weather worksheet in the gcg workbook is given the name *Serdang* (Fig. 6.5).

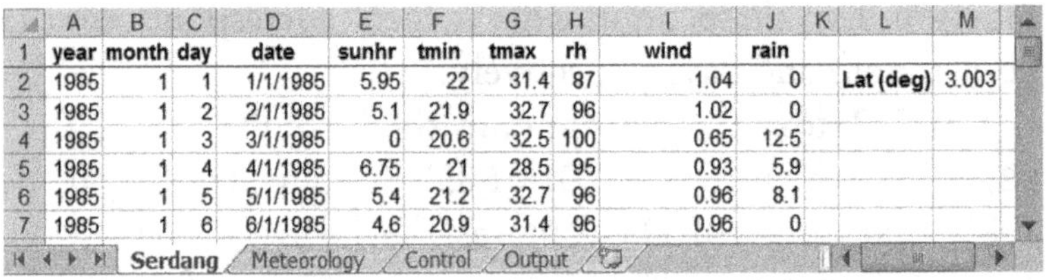

	A	B	C	D	E	F	G	H	I	J	K	L	M
1	year	month	day	date	sunhr	tmin	tmax	rh	wind	rain			
2	1985	1	1	1/1/1985	5.95	22	31.4	87	1.04	0		Lat (deg)	3.003
3	1985	1	2	2/1/1985	5.1	21.9	32.7	96	1.02	0			
4	1985	1	3	3/1/1985	0	20.6	32.5	100	0.65	12.5			
5	1985	1	4	4/1/1985	6.75	21	28.5	95	0.93	5.9			
6	1985	1	5	5/1/1985	5.4	21.2	32.7	96	0.96	8.1			
7	1985	1	6	6/1/1985	4.6	20.9	31.4	96	0.96	0			

`H  ◀  ▶  H   Serdang ╱ Meteorology ╱ Control ╱ Output ╱`

Fig. 6.5. The Serdang weather data stored in a worksheet named *Serdang* in the gcg workbook.

You can also include the weather files for other sites, but ensure these weather files are prepared according to the format as previously discussed for the Serdang weather data and that the weather data for each site is stored in a separate worksheet (Fig. 6.6).

158

	A	B	C	D	E	F	G	H	I	J	K	L	M
1	year	month	day	date	sunhr	tmax	tmin	rh	wind	rain			
2	1990	1	1	1/1/1990	0	2.2	-0.4	100	1.3	0		Lat (deg)	52
3	1990	1	2	2/1/1990	0.4	2.2	-0.4	92.03	3.4	0			
4	1990	1	3	3/1/1990	0	2.4	0	93.06	2.9	0			
5	1990	1	4	4/1/1990	0.1	1.6	-1.2	88.75	2.2	0.1			
6	1990	1	5	5/1/1990	0	3.7	1.6	93.35	4.1	0			
7	1990	1	6	6/1/1990	3.5	6.2	2.4	80.67	3.2	0.7			

| ◄ ◄ ► ►| **Netherlands** / Serdang / Meteorology / Control / Output / ... |

Fig. 6.6. An example where the Netherlands weather data are included in the gcg model. This weather data must exist as a separate worksheet (named *Netherlands* in this case) and formatted following that in the Serdang worksheet.

6.4.3 Implementation

Eq. 6.1 to 6.24 will be implemented in the Meteorology worksheet. Many cells will given cell names (Table 6.2) to make reading and understanding the various formulas much easier.

Table 6.2. Cell names defined for the meteorology model component (BuildIt cell names are listed first).

Worksheet	Cell	Cell name
Control	B6	_stepsize
	B7	_step
	B8	_criteria
Output	A3	_read
	A8	_write
Meteorology	B2	site
	B5	p
	B6	Tdmax
	B9	SB
	E2	lat
	E6	sunhr
	E7	Tmin
	E8	Tmax

159

Worksheet	Cell	Cell name
	E9	Tmean
	E10	RHd
	E11	ud
	E12	Pg
	E15	tsr
	E16	tss
	E17	DL
	E20	decl
	E21	ha
	E25	suninc
	E26	sunhgt
	E29	sunazi
	E32	Tset
	E36	Ta
	E37	Tak
	E40	Tdcal
	E41	es
	E42	ea
	E43	vpd
	E44	slopesvp
	E47	RH
	E50	umin
	E51	umax
	E53	u
	H2	Ic
	H4	Ietd
	H5	Itd
	H15	Idfd
	H16	Idrd
	H18	Iet
	H24	It
	H36	Idf

Worksheet	Cell	Cell name
	H37	Idr
	H40	RnL
	H41	Rn
Control	E2	date
	E3	doy
	E4	th

Fig. 6.7 to 6.13 show the various parts of the Meteorology worksheet.

	A	B	C
1	INPUT		
2	site	Serdang	
3	Angstrom, b_0	0.29	
4	Angstrom, b_1	0.42	
5	surface albedo, p	0.15	
6	$T_{d(max)}$	25	
7			
8	CONSTANTS		
9	Stefan-Boltzmann	=5.67*10^-8	
10			

Fig. 6.7. Meteorology worksheet: model inputs.

Model inputs are placed in one section of the Meteorology worksheet (Fig. 6.7).

Inputs required are the two Angstrom coefficients for Eq. 6.7 (cell B3 and B4), surface albedo for Eq. 6.18 (cell B5), and the maximum dew point temperature used in Eq. 6.21b (cell B6).

161

Cell B2, in particular, contains the name "Serdang" because this is the name of the worksheet in which the Serdang weather data are stored. These model inputs correspond to the meteorological properties of Serdang.

In Fig. 6.8, cell E2 contains the site latitude in radians. The function RADIANS is used to convert the latitude from degrees to radians. The site latitude is stored in cell M2 in the Serdang worksheet (Fig. 6.5).

Cell E2 could contain the following formula: "=RADIANS(Serdang!M2)" to convert 3.003 (in degrees) stored in cell M2 in the Serdang worksheet into its equivalent value in radians. However, a more flexible and elegant expression is used here instead.

	C	D	E
1			
2		latitude, λ (rad)	=RADIANS(INDIRECT(site&"!M2"))
3		Weather ref.	=site&"!D:J"
4			
5		**DAILY WEATHER**	
6		s	=VLOOKUP(date,INDIRECT($E$3),2,FALSE)
7		T_{min}	=VLOOKUP(date,INDIRECT($E$3),3,FALSE)
8		T_{max}	=VLOOKUP(date,INDIRECT($E$3),4,FALSE)
9		T_{mean}	=AVERAGE(E7:E8)
10		RH_d	=VLOOKUP(date,INDIRECT($E$3),5,FALSE)
11		u_d	=VLOOKUP(date,INDIRECT($E$3),6,FALSE)
12		P_g	=VLOOKUP(date,INDIRECT($E$3),7,FALSE)
13			

Fig. 6.8. Meteorology worksheet: reading the weather data to obtain the meteorological properties for the given date.

	C	D	E
14		**SUNRISE/SET**	
15		t_{sr}	=24-E16
16		t_{ss}	=12+(12/PI())*ACOS(-E22/E23)
17		DL	=E16-E15
18			
19		**SOLAR POSITION**	
20		δ	=-0.4093*COS(2*PI()*(doy+10)/365)
21		τ	=PI()/12*(th-12)
22		a	=SIN(decl)*SIN(lat)
23		b	=COS(decl)*COS(lat)
24		a/b	=E22/E23
25		θ (rad)	=ACOS(E22+E23*COS(ha))
26		β (rad)	=(PI()/2)-suninc
27		n1	=(SIN(lat)*SIN(sunhgt)-SIN(decl))/ (COS(lat)*COS(sunhgt))
28		n2	=ACOS(MAX(-1,MIN(1,E27)))
29		ϕ (rad)	=PI()+IF(th<12,-E28,E28)
30			

Fig. 6.9. Meteorology worksheet: calculations for sunrise and sunset hours, daylength, and solar position.

	D	E
31	**AIR TEMP.**	
32	T_{set}	=Tmin+(Tmax-Tmin)*SIN((PI()*(tss-tsr-1.5))/DL)
33	T_a (1)	=Tset+((Tmin-Tset)*(th+tsr))/((tsr+1.5)+tsr)
34	T_a (2)	=Tset+((Tmin-Tset)*(th-tss))/((tsr+1.5)+tsr)
35	T_a (3)	=Tmin+(Tmax-Tmin)*SIN((PI()*(th-tsr-1.5))/DL)
36	T_a (°C)	=IF(th<(tsr+1.5),E33,IF(th>tss,E34,E35))
37	T_a (K)	=Ta+273.15
38		

Fig. 6.10. Meteorology worksheet: air temperature.

	D	E
39	VAP. PRESS.	
40	$T_{d(cal)}$	=MIN(Ta,Tdmax)
41	$e_s[T_a]$	=6.1078*EXP(17.269*Ta/(Ta+237.3))
42	e_a	=6.1078*EXP(17.269*Tdcal/(Tdcal+237.3))
43	D	=es-ea
44	Δ	=(25029.4*EXP(17.269*Ta/(Ta+237.3)))/(Ta+237.3)^2
45		
46	HUMIDITY	
47	RH	=100*ea/es
48		
49	WIND SPEED	
50	u_{min}	=0.5*ud
51	u_{max}	=umin-(12*PI()*(umin-ud))/DL
52	Δu	=umax-umin
53	u	=umin+IF(OR(th<tsr+1.5,th>tss+1.5),0, E52*SIN(PI()*(th-tsr-1.5)/DL))

Fig. 6.11. Meteorology worksheet: air vapor pressure, relative humidity, and wind speed.

	F	G	H
1		SOLAR RADIATION	
2		I_c	=1370*(1+0.033*COS(2*PI()*(doy-10)/365))
3		n3	=E22*ACOS(-E24)+E23*SQRT(1-(E24)^2)
4		$I_{et,d}$	=3600*Ic*24/PI()*H3
5		$I_{t,d}$	=Ietd*(B3+B4*sunhr/DL)
6			
7		$I_{t,d}$ / $I_{et,d}$	=Itd/Ietd
8		$I_{df,d}$ (1)	=Itd
9		$I_{df,d}$ (2)	=Itd*(1-2.3*(H7-0.07)^2)
10		$I_{df,d}$ (3)	=Itd*(1.33-1.46*H7)
11		$I_{df,d}$ (4)	=Itd*0.23
12		case (1) true?	=H7<0.07
13		case (2) true?	=AND(H7>=0.07,H7<0.35)
14		case (3) true?	=AND(H7>=0.35,H7<0.75)
15		$I_{df,d}$	=IF(H12,H8,IF(H13,H9,IF(H14,H10,H11)))
16		$I_{dr,d}$	=Itd-Idfd
17			

Fig. 6.12. Meteorology worksheet: daily solar radiation.

	F	G	H
18		I_{et}	=MAX(0,Ic*SIN(sunhgt))
19		n4	=PI()*Itd/86400
20		n5	=E22*ACOS(-E24)+E23*SQRT(1-E24^2)
21		ψ	=H19/H20
22		aψ	=E22*H21
23		bψ	=E23*H21
24		I_t	=MAX(0,H22-H23*COS(PI()*th/12))
25			
26		I_t / I_{et}	=IF(Iet<=0,0,It/Iet)
27		R	=0.847-1.61*SIN(sunhgt)+1.04*SIN(sunhgt)^2
28		K	=(1.47-H27)/1.66
29		I_{df} (1)	=It
30		I_{df} (2)	=It*(1-6.4*(H26-0.22)^2)
31		I_{df} (3)	=It*(1.47-1.66*H26)
32		I_{df} (4)	=It*H27
33		case (1) true?	=H26<=0.22
34		case (2) true?	=AND(H26>0.22,H26<=0.35)
35		case (3) true?	=AND(H26>0.35,H26<=H28)
36		I_{df}	=IF(H33,H29,IF(H34,H30,IF(H35,H31,H32)))
37		I_{dr}	=It-Idf
38			
39		**NET RADIATION**	
40		R_{nL}	=0.98*SB*Tak^4*(1.31*(ea/Tak)^(1/7)-1)
41		R_n	=(1-p)*It+RnL

Fig. 6.13. Meteorology worksheet: hourly solar radiation and net radiation.

In cell E2 (Fig. 6.8), the expression "`site&"!M2"`" concatenates the text held in `site` (cell B2; Table 6.2) with the text "`!M2`" to produce the combined text "Serdang!M2". The ampersand symbol `&` joins the two texts: "Serdang" (the text in cell B2) and "!M2". However, if we had typed in cell E2: "`=RADIANS(site&"!M2")`", it would have resulted in an error because the expression "`site&"!M2"`" is literally taken as a cell reference. Although "Serdang!M2" is a valid cell reference, the literal expression "`site&"!M2"`" is not.

So, to convert a text into a cell reference, we need to use the Excel function `INDIRECT`. The Excel function `INDIRECT` returns the cell reference specified by a text so that "`=RADIANS(INDIRECT(site&"!M2"))`" is as if we had typed: "`=RADIANS(Serdang!M2)`".

We will continue to use this technique to lookup the weather data. In cell E3 (Fig. 6.8), the expression "`=site&"!D:J"`" evaluates to "Serdang!D:J". Cells E6:E8 and E10:E12 use the combination of the `VLOOKUP` and `INDIRECT` functions to lookup the meteorological properties for the given date.

The `VLOOKUP` Excel function has the following specification:

```
VLOOKUP( value,
         table_array,
         index_number,
         range_lookup)
```

which searches for `value` in the leftmost column of `table_array` and returns the value on the same row based on the `index_number`. The last argument `range_lookup` is set to FALSE if an exact match is required, else set to TRUE to return the next largest value that is less than value if an exact match cannot be found.

For instance, to lookup the sunshine hours for 12 February 2000 for Serdang, we would write:

VLOOKUP(DATE(2000,2,12),Serdang!D:J,2,FALSE)

where the first argument is the DATE function which returns 12 Feb. 2000. The second argument refers to the table in the Serdang worksheet in columns D to J (*i.e.*, Serdang!D:J).

Refer to the Serdang worksheet (Fig. 6.5), and we would find that the entire weather data (from 1985 to 2004) stretches from cell A2 to J7306. However, the first three columns (A, B and C) store the year, month and day, respectively, and they will not be used in weather lookup. Instead we will use the fourth column (D) onwards to lookup the weather data. Consequently, the table for weather lookup actually begins from column D to J. And since the entire Serdang worksheet is dedicated to storing the weather data, we can eliminate the row numbers and succinctly write "Serdang!D:J" instead of "Serdang!D2:J7306".

In our example, once the date 12 February 2000 is located from the table, the value from the second column (as specified by the third argument, index_number) of the table is returned. Looking in the Serdang worksheet again (Fig. 6.5), we see that the sunshine hours data are stored in column E which is the second column after column D. The last argument is set to FALSE because an exact match (not an approximate match) must be found.

As previously stated, cell E3 contains the formula: "=site&"!D:J"" which gives "Serdang!D:J". But if we write the following lookup formula:

VLOOKUP(DATE(2000,2,12),E3,2,FALSE)

fails because the second argument states that cell E3 is where the lookup table is located. Cell E3 contains the reference where the weather table is located and is not the lookup table itself. To circumvent this problem, we will again use the INDIRECT function. This function returns the reference to a cell based on its text representation, so writing the formula as

VLOOKUP(DATE(2000,2,12),INDIRECT(E3),2,FALSE)

is as if we had written:

VLOOKUP(DATE(2000,2,12),Serdang!D:J,2,FALSE)

Consequently, for the current simulation date, daily weather properties will be looked up from the weather table via the function

VLOOKUP(date,INDIRECT(E3),*index_number*,FALSE)

where *index_number* is 2 to 7, depending on the weather property, and date is the cell name referring to cell E2 in the Control worksheet (Table 6.2) which holds the current simulation date.

The use of the INDIRECT function is very convenient because if we wish to use the weather data of another site, we only need to change cell B2. For instance, if we wish to use the Netherlands weather data (Fig. 6.6), we only need to enter "Netherlands" (the name of the worksheet where the Netherlands data are kept) in cell B2.

Lastly for Fig. 6.8, cell E9 determines the average of the minimum and maximum air temperature.

In Fig. 6.9, cells E15:E17 implement Eq. 6.2 to 6.4 and cells E20:E29 Eq. 6.1 and 6.10 to 6.12. Calculations for the air temperature are carried out in cells E32:37 (Fig. 6.10) based on Eq. 6.17. Cell E36 uses the logical IF function to decide which of the three equations (cells E33, 34, or 35) to use, depending on the current hour. Note that cell E37 converts the air temperature from °C to K which would be later used in the net radiation calculations.

Fig. 6.11 shows the calculations for the air vapor pressure, relative humidity, and wind speed based on Eq. 6.19 to 6.24. Cell E40 implement Eq. 6.21b.

Calculations for the daily solar radiation are performed in cells H2:H16 (Fig. 6.12) based on Eq. 6.5 to 6.9. Cells H8:H11 implement the four sub-equations in Eq. 6.8, and the logical IF function is used in cell H36 to determine which of these four sub-equation is to be used to determine the diffuse solar radiation $I_{df,d}$.

Cells H18:37 (Fig. 6.13) carry out the calculations to determine the hourly solar radiation based on Eq. 6.13 to 6.16. Cells H18 and H24 use the MAX function to ensure the calculated I_{et} and I_t never have negative values which would be invalid. Cells H29:32 implement the four sub-equations in Eq. 6.15, and, depending on the I_t/I_{et} value (cell H26), the logical IF function in cell H36 decides which of these four sub-equations will be used to calculate the diffuse solar radiation. Note that cell H26 uses the IF function to account for the situation where the denominator (I_{et}) is 0. Lastly, the net radiation is calculated in cells H40:41 based on Eq. 6.18.

The implementation of the meteorology component of the gcg model is nearly complete at this point. Let us now simulate the hourly values of various meteorological properties for any given date.

The Control worksheet contains information on the loop (Fig. 6.14). Cells B5:B8 instruct BuildIt to run the simulation for t_h (cell E4) = 0, 1, 2, ..., 24.

Notice that the input date in cell B2 is specified as "=DATE(2000,9,10)" to mean 10 Sept. 2000. You should enter the date via the DATE function rather than entering "10/09/00", as this format can be interpreted differently on different computer systems.

	A	B	C	D	E
1	INPUT				
2	date	=DATE(2000,9,10)		current date	=B2
3				doy	=date-DATE(YEAR(date),1,0)
4	CONTROL			hour	=_step
5	maxsteps	24			
6	stepsize	1			
7	step				
8	criteria	=th<=B5			
9					

Fig. 6.14. Control worksheet: setup to run at hourly steps from 0:00 to 24:00 hours on 10 Sept. 2000.

Cell E2 (defined with cell name date; Table 6.2) refers to the current simulation date, set equal to cell B2. In cell E3, the formula to determine the day of year (doy) for the current date is "=date-DATE(YEAR(date),1,0)" which gives the number of days from the first day of the same year (1 January). However, since the day of year starts as 1 (not 0), the subtraction of dates must be specified as "=date-DATE(YEAR(date),1,0)" and not as "=B2-DATE(YEAR(date),1,1)".

It might seem odd that the day argument in DATE is given as 0, but doing so will ensure that the day of year for 1 January is 1 (not 0). We also use the YEAR function to extract the year for the current simulation date, so that the day of year will always be within the range of 1 and 365 (or 366 for leap years).

Note that subtracting the two dates in cell E3 would produce the day of year to be displayed in the date format (*e.g.*, displayed confusingly as "10-09-00"). We need to change the display format of this cell to *General*. Right click on cell E3, and choose "Format Cells" from the popup menu. Choose the "Number" tab, then set the display format to "General" (Fig. 6.15).

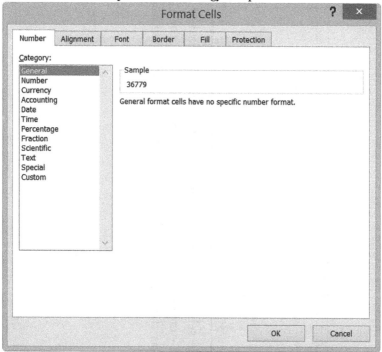

Fig. 6.15. Format Cells dialog: change how the day of year is displayed by changing the display format to General.

The parameters and meteorological properties we wish to see in the model output are as shown in Fig. 6.16. Cell A3 in the Output worksheet is defined with the cell name _read (Table 6.2), so the parameters and meteorological properties listed from cell A3 to N3 will be displayed in the output listing which would start from cell A8 (defined with cell name _write; Table 6.2) in the same Output worksheet. Note that cells B3 and C3 use the function DEGREES to convert the solar inclination and solar azimuth values from unit radians to degrees.

	A	B	C	D	E	F	G	H	I	J	K	L	M	N	O
1	TO OUTPUT														
2	th	θ	φ	T_a	RH	$e_s[T_a]$	e_a	D	u	l_{et}	l_t	l_{df}	l_{dr}	R_n	
3	=th	=DEGREES(suninc)	=DEGREES(sunazi)	=Ta	=RH	=es	=ea	=vpd	=u	=let	=lt	=ldf	=ldr	=Rn	
4															
5															
6															
7	OUTPUT														
8															
9															

Fig. 6.16. Output worksheet: setup to include selected model parameters and meteorological properties in the model output. The what-to-output section begins at cell A3 (defined with _read cell name) and ends at cell N3, whereas the output section begins at cell A8 (defined with _write cell name).

Once completed, choose "Start Simulation" from the BuildIt menu. The simulation results are the hourly meteorological properties for 10 Sept. 2000 from 0:00 to 24:00 hours (Fig. 6.17), and several charts are drawn based on the output listing (Fig. 6.18).

	A	B	C	D	E	F	G	H	I	J	K	L	M	N
1	TO OUTPUT													
2	th	θ	ϕ	T_a	RH	$e_s[T_a]$	e_a	D	u	I_{et}	I_t	I_{df}	I_{dr}	R_n
3	25	163	295	24.1	100	30	30	0	0.3	0	0	0	0	-24
4														
5														
6														
7	OUTPUT													
8	0	173	0	24.4	100	30.5	30.5	0	0.3	0	0	0	0	-23
9	1	163	65.3	24.1	100	30	30	0	0.3	0	0	0	0	-24
10	2	149	77.2	23.8	100	29.5	29.5	0	0.3	0	0	0	0	-25
11	3	134	81.5	23.5	100	29	29	0	0.3	0	0	0	0	-26
12	4	120	83.8	23.3	100	28.6	28.6	0	0.3	0	0	0	0	-27
13	5	105	85.2	23	100	28.1	28.1	0	0.3	0	0	0	0	-28
14	6	89.8	86.1	22.7	100	27.6	27.6	0	0.3	4.82	2.36	1.98	0.37	-26
15	7	74.8	86.8	22.4	100	27.1	27.1	0	0.3	352	172	113	58.9	117
16	8	59.9	87.2	23.6	100	29.1	29.1	0	0.43	676	331	218	113	255
17	9	44.9	87.5	26.1	93.8	33.8	31.7	2.08	0.66	954	467	307	160	375
18	10	30	87.4	28.3	82.4	38.4	31.7	6.75	0.87	1168	571	376	195	462
19	11	15	86.1	30.1	74.3	42.6	31.7	11	1.05	1302	637	419	218	517
20	12	0.91	360	31.4	69.1	45.8	31.7	14.2	1.17	1348	659	434	225	535
21	13	15	274	32	66.6	47.6	31.7	15.9	1.23	1302	637	419	218	516
22	14	30	273	32	66.6	47.6	31.7	15.9	1.23	1168	571	376	195	460
23	15	44.9	273	31.4	69.1	45.8	31.7	14.2	1.17	954	467	307	160	372
24	16	59.9	273	30.1	74.3	42.6	31.7	11	1.05	676	331	218	113	257
25	17	74.8	273	28.3	82.4	38.4	31.7	6.75	0.87	352	172	113	58.9	123
26	18	89.8	274	26.1	93.8	33.8	31.7	2.08	0.66	4.82	2.36	1.98	0.37	-20
27	19	105	275	25.8	95.5	33.2	31.7	1.48	0.43	0	0	0	0	-22
28	20	120	276	25.5	97.1	32.6	31.7	0.94	0.3	0	0	0	0	-22
29	21	134	278	25.2	98.7	32.1	31.7	0.4	0.3	0	0	0	0	-22
30	22	149	283	24.9	100	31.6	31.6	0	0.3	0	0	0	0	-22
31	23	163	295	24.7	100	31	31	0	0.3	0	0	0	0	-23
32	24	173	360	24.4	100	30.5	30.5	0	0.3	0	0	0	0	-23
33														

Fig. 6.17. Output worksheet: simulation of the hourly meteorological properties for Serdang (on 10 Sept. 2000).

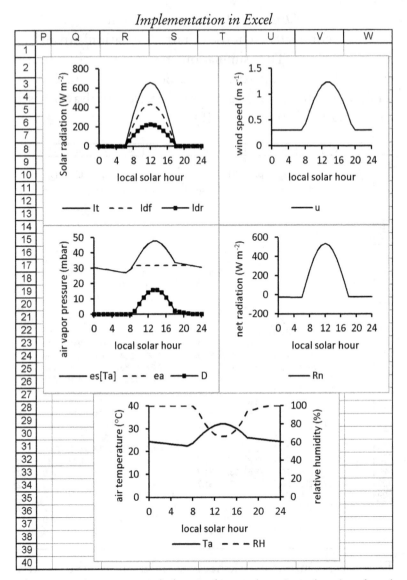

Fig. 6.18. Output worksheet: charts showing the simulated hourly change for several meteorological properties for Serdang (simulations are for 10 Sept. 2000).

6.5 Exercises

1. Using the Serdang weather data, run the model for different dates, and compare their similarities and differences between their meteorological properties. The dates are as follows:
   a) 10 Mar. 2000
   b) 10 Jun. 2000
   c) 10 Dec. 2000

2. Obtain the weather data for one or more years for your site, and run the model to simulate the hourly meteorological properties for a selected date.
   Note: using the weather data of another site should not require you to modify the model but merely to add a new worksheet containing the weather data into the workbook. The weather data for your site should be tabulated following the arrangement and format as in the Serdang worksheet (*see* Fig. 6.5 and 6.6).

3. Haze is an atmospheric condition where suspended dust, smoke, or other dry particles cause a significant reduction in sky clarity. Haze can lower incoming solar irradiance and even air temperature.
   Assume Serdang had experienced haze on 10 Sept. 2000. Without modifying the tabulated weather data in the Serdang worksheet, modify the model and simulate the hourly meteorological properties for Serdang on 10 Sept. 2000 if the haze had reduced the sunshine hours for that day by 2 hours and lowered both the minimum and maximum air temperatures by 1 °C.
   Compare Serdang's hourly meteorological properties due to haze with that without haze. Plot charts to help in the comparisons.

4. Modify the model so that sunshine hours, s, is not read from the tabulated weather data but is calculated by:

$$s = \bar{s} + A\sin\left[\frac{2\pi}{364}\left(t_d - t_{peak} + 91\right)\right]$$

where $\bar{s}$ is the annual average sunshine hours; t_d is the day of year; A is the amplitude of the sine curve (the difference in hours between the maximum annual s and $\bar{s}$); and t_{peak} is the day of year when s is the highest. Simulate Serdang's hourly meteorological properties on 10 Sept. 2000, taking $\bar{s}$, A, and t_{peak} as 4 hours, 1.6 hours, and 118, respectively.

Chapter 7. Canopy photosynthesis component

The canopy photosynthesis model component calculates the amount of CO_2 assimilated by the canopies. The following equations are taken (some adapted) from Teh et al. (2004), Campbell and Norman (1998), Goudriaan and van Laar (1994), and Goudriaan (1977).

The various properties calculated by the canopy photosynthesis component is visually depicted in Fig. 7.1.

7.1   Equations

7.1.1   PAR (photosynthetically active radiation) components within canopies

Four flux components exist within canopies, and they can be described by

$$Q_{p,dr} = (1 - p_p)Q_{dr} \exp(-k_{dr}L) \tag{7.1}$$

$$Q_{p,dr,dr} = (1 - p_p)Q_{dr} \exp\left(-\sqrt{\alpha}k_{dr}L\right) \tag{7.2}$$

$$Q_{p,dr,\alpha} = \frac{1}{2}\left(Q_{p,dr} - Q_{p,dr,dr}\right) \tag{7.3}$$

$$\bar{Q}_{p,df} = \frac{(1 - p_p)Q_{df}\left[1 - \exp\left(-\sqrt{\alpha}k_{df}L\right)\right]}{\sqrt{\alpha}k_{df}L} \tag{7.4}$$

where $Q_{p,dr}$ is the PAR for unintercepted beam *with* scattering (μmol photons m^{-2} ground s^{-1}); $Q_{p,dr,dr}$ is the PAR irradiance for unintercepted beam *without* scattering (μmol photons m^{-2} ground s^{-1}); $Q_{p,dr,\alpha}$ is the PAR irradiance of the scattered component *only* (μmol photons m^{-2} leaf s^{-1}); and $\bar{Q}_{p,df}$ is the mean diffuse irradiance (μmol photons m^{-2} leaf s^{-1}). Note that both $Q_{p,dr,\alpha}$ and $\bar{Q}_{p,df}$ are based on a per unit leaf area (not per unit ground area as for the other two components: $Q_{p,dr}$ and $Q_{p,dr,dr}$).

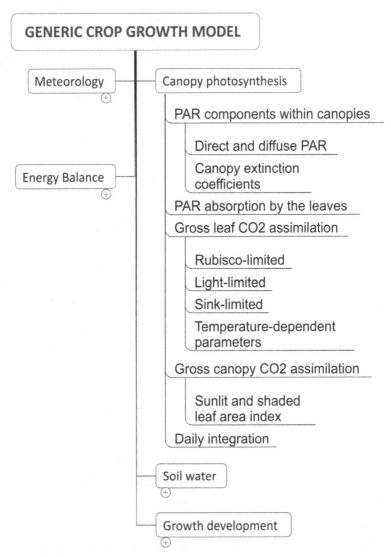

Fig. 7.1. The canopy photosynthesis model component.

Canopy reflection coefficient for PAR, p_p, is taken as 0.04, and the leaf scattering coefficient for PAR, α, is taken as 0.8.

7.1.1.1 Direct and diffuse PAR

$$Q_{dr} = I_{dr} \times 0.5 \times 4.55 \tag{7.5}$$

$$Q_{df} = I_{df} \times 0.5 \times 4.55 \tag{7.6}$$

where Q_{dr} and Q_{df} are, respectively, the hourly direct and diffuse PAR irradiance, both expressed in μmol photons m^{-2} ground s^{-1}; and I_{dr} and I_{df} are the hourly direct and diffuse solar irradiance, respectively (W m^{-2} ground).

Note: total PAR irradiance is assumed to be 50% of total solar irradiance. Further multiplication by 4.55 is to convert unit W m^{-2} ground to μmol photons m^{-2} ground s^{-1}.

7.1.1.2 Canopy extinction coefficient for direct PAR

It is assumed the leaves in the canopies are randomly distributed, so that

$$k_{dr} = \frac{\omega}{2\cos\theta} \tag{7.7a}$$

$$\omega = \omega_0^{1-2\theta/\pi} \tag{7.7b}$$

$$\omega_0 = -\frac{2\cos\theta}{L}\ln\left\{\tau_b + (1-\tau_b)\exp\left[-\frac{L}{2(1-\tau_b)\cos\theta}\right]\right\} \tag{7.7c}$$

$$\tau_b = \begin{cases} 1 - L/L_{\max} & L < L_{\max} \\ 0 & L \geq L_{\max} \end{cases} \tag{7.7d}$$

where k_{dr} is the canopy extinction coefficient for direct PAR; θ is the solar inclination (radians); ω is the canopy clustering coefficient; ω_0 is the canopy clustering coefficient when the sun is at zenith (highest point in the sky); L is the total leaf area index (m² leaf m⁻² ground); L_{max} is the maximum leaf area index for full canopy cover (taken as 3 m² m⁻²); and τ_b is the canopy gap fraction.

7.1.1.3 Canopy extinction coefficient for diffuse PAR

$$k_{df} = \frac{a + c \cdot L + e \cdot L^2 + g \cdot L^3 + i \cdot L^4 + k \cdot L^5}{1 + b \cdot L + d \cdot L^2 + f \cdot L^3 + h \cdot L^4 + j \cdot L^5} \qquad (7.8)$$

where k_{df} is the canopy extinction coefficient for diffuse PAR; and L is the total leaf area index (LAI) (m⁻² leaf m⁻² ground). The values for the coefficients are from Table 7.1.

Table 7.1. Coefficient values (in nine decimal places) for the canopy extinction coefficient for diffuse PAR (Eq. 7.8).

Coefficient	Value	Coefficient	Value
a	1.002331825	*g*	0.030599924
b	2.861841206	*h*	0.015715751
c	2.122702035	*i*	0.000491650
d	0.093153495	*j*	0.000471880
e	0.477011110	*k*	0.000000694
f	0.148586050		

Note: k_{df} has already been corrected for canopy clustering. Like that for k_{dr}, the assumption in k_{df} is that the leaves are also randomly distributed in the canopies.

7.1.2   PAR absorption by the leaves

$$Q_{sl} = \alpha\left(k_{dr}Q_{dr} + \overline{Q}_{p,df} + Q_{p,dr,\alpha}\right) \qquad (7.9)$$

$$Q_{sh} = \alpha\left(\overline{Q}_{p,df} + Q_{p,dr,\alpha}\right) \qquad (7.10)$$

where Q_{sl} and Q_{sh} are the PAR flux densities absorbed by the sunlit leaves and shaded leaves, respectively (μmol photons m^{-2} leaf s^{-1}); k_{dr} is the canopy extinction coefficient for direct PAR; Q_{dr} is the hourly direct PAR irradiance (μmol photons m^{-2} ground s^{-1}); $Q_{p,dr,\alpha}$ is the PAR irradiance of the scattered component *only* (μmol photons m^{-2} leaf s^{-1}); $\overline{Q}_{p,df}$ is the mean diffuse irradiance (μmol photons m^{-2} leaf s^{-1}); and α is the leaf scattering coefficient (where in this case, it is also the leaf absorption coefficient) for PAR (0.8).

Note: PAR flux densities absorbed (Q_{sl} and Q_{sh}) are based on a per unit leaf area (not per unit ground area).

7.1.3   Gross leaf CO_2 assimilation

$$\Lambda_{sl/sh} = MIN\left(v_c, v_{q,sl/sh}, v_s\right) \qquad (7.11)$$

where $\Lambda_{sl/sh}$ is the gross leaf CO_2 assimilation (μmol CO_2 m^{-2} leaf s^{-1}) for either sunlit (subscript *sl*) or shaded (subscript *sh*) leaves; and MIN is a function to select the minimum of the three assimilation rates (v_c, $v_{q,sl/sh}$ and v_s). In other words, gross leaf CO_2 assimilation is limited by the most limiting factor of either v_c, $v_{q,sl/sh}$, or v_s.

7.1.3.1 Rubisco-limited assimilation (v_c)

$$v_c = \frac{V_{c,max}(C_i - \Gamma^*)}{K_c(1 + O_a/K_o) + C_i}$$
(7.12)

where v_c is the Rubisco-limited leaf CO_2 assimilation rate (μmol CO_2 m^{-2} leaf s^{-1}); K_c is the Michaelis-Menten constant for CO_2 (μmol mol^{-1}); K_o is the Michaelis-Menten constant for CO_2 (μmol mol^{-1}); O_a is the ambient O_2 concentration in air (210000 μmol mol^{-1}); C_i is the intercellular CO_2 concentration (270 μmol mol^{-1}); $V_{c,max}$ is the Rubisco capacity rate (μmol m^{-2} leaf s^{-1}); and Γ^* is the CO_2 compensation point (μmol mol^{-1}), determined by

$$\Gamma^* = O_a/2\tau$$
(7.13)

where τ is CO_2/O_2 specificity factor (μmol μmol^{-1}). The values for K_c, K_o, τ, and $V_{c,max}$ are listed in Table 7.2.

Table 7.2. Parameter values for the photosynthesis calculations. Parameters with subscript (25) denote their values at 25 °C.

$\xi_{(25)}$	Description	Value	Q_{10}
$K_{c(25)}$	Michaelis-Menten constant for CO_2 (μmol mol^{-1})	300	2.1
$K_{o(25)}$	Michaelis-Menten constant for O_2 (μmol mol^{-1})	300000	1.2
$\tau_{(25)}$	CO_2 / O_2 specificity factor (μmol μmol^{-1})	2600	0.57
$V_{c,max(25)}$	Rubisco capacity (μmol m^{-2} s^{-1})	200	2.4

Note: K_c, K_o, τ, and $V_{c,max}$ must be corrected for canopy temperature (*see* section 7.1.3.4).

7.1.3.2 Light-limited assimilation (v_q)

$$v_{q,sl/sh} = e_m \alpha Q_{sl/sh} \frac{C_i - \Gamma^*}{C_i + 2\Gamma^*} \qquad (7.14)$$

where $v_{q,sl/sh}$ is the light-limited leaf CO_2 assimilation rate (μmol CO_2 m^{-2} leaf s^{-1}) for either sunlit (subscript *sl*) or shaded (subscript *sh*) leaves; Γ^* is the CO_2 compensation point (μmol mol^{-1}); C_i is the intercellular CO_2 concentration (270 μmol mol^{-1}); $Q_{sl/sh}$ is the PAR absorbed by either sunlit (subscript *sl*) or shaded (subscript *sh*) leaves (μmol m^{-2} leaf area s^{-1}); α is the leaf absorption of PAR (0.8); and e_m is the intrinsic quantum efficiency or quantum yield (0.08).

Note: Γ^* must be corrected for canopy temperature (see section 7.1.3.4).

7.1.3.3 Sink-limited assimilation (v_s)

$$v_s = \frac{V_{c,max}}{2} \qquad (7.15)$$

where v_s is the sink-limited leaf CO_2 assimilation rate (μmol CO_2 m^{-2} leaf s^{-1}); and $V_{c,max}$ is the Rubisco capacity rate (μmol m^{-2} leaf s^{-1}).

Note: $V_{c,max}$ must be corrected for canopy temperature (*see* section 7.1.3.4).

7.1.3.4 Temperature-dependent parameters

Parameters K_c, K_o, τ, and $V_{c,max}$ are sensitive to the canopy or foliage temperature T_f (°C). Consequently, their values are adjusted according to

$$\xi = \xi_{(25)} Q_{10}^{(T_f - 25)/10}$$

where $\xi_{(25)}$ is the parameter value at 25 °C ($K_{c(25)}$, $K_{o(25)}$, $\tau_{(25)}$ or $V_{c,max(25)}$); T_f is the canopy temperature (°C); and Q_{10} is the relative change in the parameter ξ for every 10 °C change.

Values for the model parameters at 25 °C and their respective Q_{10} values are given in Table 7.2.

Michaelis-Menten constant for CO2

$$K_c = 300 \times 2.1^{(T_f - 25)/10} \tag{7.16}$$

where K_c is the Michaelis-Menten constant for CO_2 (μmol mol^{-1}); and T_f is the canopy temperature (°C).

Michaelis-Menten constant for O2

$$K_o = 300000 \times 1.2^{(T_f - 25)/10} \tag{7.17}$$

where K_o is the Michaelis-Menten constant for CO_2 (μmol mol^{-1}); and T_f is the canopy temperature (°C).

CO2/O2 specificity factor

$$\tau = 2600 \times 0.57^{(T_f - 25)/10} \tag{7.18}$$

where τ is the CO_2/O_2 specificity factor (μmol μmol^{-1}); and T_f is the canopy temperature (°C).

Rubisco capacity rate

$$V_{c,\max} = \frac{200 \times 2.4^{(T_f - 25)/10}}{1 + \exp\left[0.29(T_f - 40)\right]} \tag{7.19}$$

where $V_{c,max}$ is the Rubisco capacity rate (μmol m^{-2} leaf s^{-1}); and T_f is the canopy temperature (°C).

Note: $V_{c,max}$ is corrected additionally for temperatures greater than 40 °C after which causes a rapid decline in the capacity rate due to Rubisco degradation.

7.1.4   Gross canopy CO2 assimilation

$$\Lambda_{canopy} = \Lambda_{sl} L_{sl} + \Lambda_{sh} L_{sh} \tag{7.20}$$

where Λ_{canopy} is the gross canopy CO2 assimilation (μmol m^{-2} ground s^{-1}); Λ_{sl} and Λ_{sh} are the gross leaf CO2 assimilation rates for sunlit and shaded leaves, respectively (μmol m^{-2} leaf s^{-1}); and L_{sl} and L_{sh} are the sunlit and shaded leaf area index, respectively (m^2 leaf m^{-2} ground), determined by

$$L_{sl} = \frac{1 - \exp(-k_{dr} L)}{k_{dr}} \tag{7.21}$$

$$L_{sh} = L - L_{sl} \tag{7.22}$$

where L_{sl} and L_{sh} are sunlit and shaded leaf area index, respectively (m^2 leaf m^{-2} ground); L is the total leaf area index (m^2 leaf m^{-2} ground); and k_{dr} is the canopy extinction coefficient for direct PAR.

7.1.5   Daily integration

The total amount of CO_2 assimilated for the whole day is determined by integrating Eq. 7.20 over the period from sunrise (t_{sr}) to sunset (t_{ss}):

$$\Lambda_{canopy,d} = 3600 \times 30 \times 10^{-6} \times \int_{t_{sr}}^{t_{ss}} \Lambda_{canopy}\ dt \qquad (7.23)$$

where multiplication by 3600 and 30 x 10^{-6} is to convert the daily gross canopy photosynthesis, $\Lambda_{canopy,d}$, to g CH_2O m^{-2} ground day^{-1}.

7.2   Measured parameters

The following are parameters that must be measured or supplied to the canopy photosynthesis model component:

1. Canopy reflection coefficient for PAR, p_p, and leaf scattering coefficient for PAR, α. Both are taken in this book as having 0.04 and 0.8 values, respectively.
2. Maximum leaf area index for full ground cover, L_{max}, taken as 3 m^2 m^{-2}.
3. Michaelis-Menten constant for CO_2 and O_2 (K_c and K_o, respectively), CO_2/O_2 specificity factor (τ), Rubisco capacity ($V_{c,max}$). The values for these parameters at 25 °C and their corresponding Q_{10} values are taken from Table 7.2.
4. Ambient O_2 concentration in air, O_a, taken as 210000 µmol mol^{-1} and the intercellular CO_2 concentration, C_i, taken as 270 µmol mol^{-1}.
5. Quantum yield, e_m, taken as 0.08.

6. *Provisional values*: A more rigorous approach to modeling the canopy temperature (T_f) and leaf area index (L) will be done when we later implement in the energy balance and growth development model components, respectively. But at the moment, T_f and L will have the following provisional values:

    a) T_f taken as 2 °C higher than air temperature, T_a.

    b) L taken as equal to L_{max}.

7.3 Implementation

In the gcg model workbook, insert a new worksheet and name it *Photosynthesis* (Fig. 7.2). This is in addition to the four previous worksheets (Serdang, Meteorology, Control, and Output).

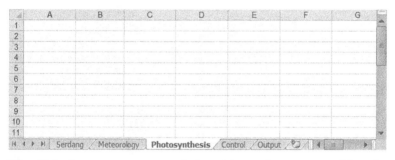

Fig. 7.2. Insert a new worksheet into the gcg workbook. This new worksheet is named *Photosynthesis* and will contain equations pertaining to canopy photosynthesis.

To improve model readability, we will define and use several more cell names as listed in Table 7.3, and we will implement Eq. 7.1 to 7.23 in the Photosynthesis worksheet as shown in Fig. 7.3 to 7.6.

Table 7.3. Cell names defined for the canopy photosynthesis model component, in addition to those already defined for the meteorology component (Table 6.2).

Worksheet	Cell	Cell name
Control	G2	_operation
Photosynthesis	B2	Lmax
	B11	Oa
	B12	Ci
	B13	pp
	B14	alpha
	B15	em
	D6	Kc
	D7	Ko
	D8	tau
	D9	Vcmax
	G3	Tf
	G4	L
	G11	kdr
	G14	kdf
	G17	Lsl
	G18	Lsh
	G21	Qdr
	G22	Qdf
	J10	Qsl
	J11	Qsh
	J14	co2pt
	J15	vc
	J16	vs
	J17	vqsl
	J18	vqsh
	J19	assim
	J21	dayassim

	A	B	C	D	E
1	INPUT				
2	L_{max}	3			
3					
4	**PARAMETERS**				
5	param.	$\xi_{(25)}$	Q_{10}	$\xi_{(Tf)}$	
6	K_c	300	2.1	=B6*C6^((Tf-25)/10)	
7	K_o	300000	1.2	=B7*C7^((Tf-25)/10)	
8	τ	2600	0.57	=B8*C8^((Tf-25)/10)	
9	$V_{c,max}$	200	2.4	=(B9*C9^((Tf-25)/10))/(1+EXP(0.29*(Tf-40)))	
10					
11	O_a	210000			
12	C_i	270			
13	P_p	0.04			
14	α	0.8			
15	e_m	0.08			
16					
17	k_{df} **coef.**				
18	*a*	1.002331825			
19	*b*	2.861841206			
20	*c*	2.122702035			
21	*d*	0.093153495			
22	*e*	0.47701111			
23	*f*	0.14858605			
24	*g*	0.030599924			
25	*h*	0.015715751			
26	*i*	0.00049165			
27	*j*	0.00047188			
28	*k*	0.000000694			
29					

Fig. 7.3. Photosynthesis worksheet: model inputs.

	E	F	G	H
1				
2		*PROVISIONAL*		
3		T_f	=Ta+2	
4		L	=Lmax	
5				
6		**EXT. COEF.**		
7		τ_b	=IF(L<Lmax,1-L/Lmax,0)	
8		$1 - \tau_b$	=1-G7	
9		ω_0	=-2*COS(suninc)/L*LN(G7+G8*EXP(-L/(2*G8*COS(suninc))))	
10		ω	=G9^(1-2*suninc/PI())	
11		k_{dr}	=MAX(0,G10/(2*COS(suninc)))	
12		n1	=B18+B20*L+B22*L^2+B24*L^3+B26*L^4+B28*L^5	
13		n2	=1+B19*L+B21*L^2+B23*L^3+B25*L^4+B27*L^5	
14		k_{df}	=G12/G13	
15				
16		**SUNLIT/SHADED**		
17		L_{sl}	=(1-EXP(-kdr*L))/kdr	
18		L_{sh}	=L-Lsl	
19				
20		**PAR ABOVE**		
21		Q_{dr}	=0.5*Idr*4.55	
22		Q_{df}	=0.5*Idf*4.55	
23				

Fig. 7.4. Photosynthesis worksheet: calculations for the canopy extinction coefficients, sunlit and shaded leaf area index, and the PAR (photosynthetically active radiation) above the canopies.

190

	H	I	J	K
1				
2		PAR CANOPIES		
3		$Q_{p,dr}$	=(1-pp)*Qdr*EXP(-kdr*L)	
4		$Q_{p,dr,dr}$	=(1-pp)*Qdr*EXP(-SQRT(alpha)*kdr*L)	
5		$Q_{p,dr,\alpha}$	=(J3-J4)/2	
6		n3	=SQRT(alpha)*kdf*L	
7		$Q_{p,df}$	=(1-pp)*Qdf*(1-EXP(-J6))/J6	
8				
9		PAR ABSORBED		
10		Q_{sl}	=alpha*(kdr*Qdr+J7+J5)	
11		Q_{sh}	=alpha*(J7+J5)	
12				
13		CO_2 ASSIM.		
14		Γ^*	=Oa/(2*tau)	
15		v_c	=Vcmax*(Ci-co2pt)/(Kc*(1+Oa/Ko)+Ci)	
16		v_s	=0.5*Vcmax	
17		$v_{q,sl}$	=em*alpha*Qsl*((Ci-co2pt)/(Ci+2*co2pt))	
18		$v_{q,sh}$	=em*alpha*Qsh*((Ci-co2pt)/(Ci+2*co2pt))	
19		Λ_{canopy}	=Lsl*MIN(vc,vs,vqsl)+Lsh*MIN(vc,vs,vqsh)	
20		$\int \Lambda_{canopy}\, dt$		
21		$\Lambda_{canopy,d}$ (g CH_2O m^{-2})	=J20*3600*30*10^-6	
22				

Fig. 7.5. Photosynthesis worksheet: calculations for the PAR (photosynthetically active radiation) fluxes and the CO_2 assimilation rates.

In Fig. 7.3, cell B3 stores the maximum leaf area index taken as 3 m^2 m^{-2}, as stated earlier. Cell B6:C9 are values of the temperature-dependent CO_2 assimilation rate parameters, as listed in Table 7.2. Cells D6:D9 calculate the values of these temperature-dependent parameters for the current canopy temperature based on Eq. 7.16 to 7.19.

191

Cells B11:B15 are the values for the non temperature-dependent assimilation parameters. Cells B18:B28 are the coefficient values for calculating the canopy extinction coefficient for diffuse radiation (Table 7.1).

In Fig. 7.4, cells G3 and G4 are the provisional values for canopy temperature and leaf area index, respectively. Cells G7:G14 implement Eq. 7.7 and 7.8 to determine the canopy extinction coefficient for direct and diffuse radiation, respectively. Cells G17 and G18 implement Eq. 7.21 and 7.22 to determine the sunlit and shaded leaf area index, respectively. Cells G21 and G22 implement Eq. 7.5 and 7.6 to determine the direct and diffuse PAR irradiance above the canopies, respectively.

In Fig. 7.5, cells J3:J7 implement Eq. 7.1 to 7.4 to determine the various PAR fluxes within the canopies. Cells J10 and J11 implement Eq. 7.9 and 7.10 to determine the amount of PAR absorbed by sunlit and shaded leaves, respectively. Cells J14:J18 calculate the leaf CO_2 assimilation rates based on Eq. 7.12 to 7.15, and cell J19 calculate the gross canopy CO_2 assimilation rate based on Eq. 7.11 and 7.20.

Integration of Eq. 7.20 over the time period of sunrise to sunset will be performed using the BuildIt ITG action (Table 3.5). We will create the operation section in the Control worksheet (Fig. 7.6).

In the Control worksheet, give cell G2 the cell name _operation (Table 7.3), so this cell G2 marks the start of the operation section. At the moment, this operation section contains only one BuildIt action. The ITG action is specified from cells G2:N2 to integrate cell Photosynthesis!J19 (assim; Table 7.3) from the time of sunrise to sunset and to store the integration result in cell Photosynthesis!J20. Note that cell M2 and N2 in the

Control worksheet are left blank, so they are TRUE by default, which means a single integration is performed and this ITG action is always executed by BuildIt (*see* section 3.2).

	F	G	H	I	J	K	L	M	N	O
1		OPERATION								
2		ITG	=th	=assim	=Photosynthesis!J20	=tsr	=tss			
3										
4										
5										

Fig. 7.6. Control worksheet: setup of the operation section to use the ITG action for numerical integration of the CO_2 assimilation rate.

Cell J21 in Fig. 7.5 multiplies the integration result with 3600 and 30 x 10^{-6} to convert the daily gross canopy photosynthesis to g CH_2O m^{-2} ground day^{-1} (Eq. 7.23).

Let us now simulate the daily assimilation rate for ten days at the Serdang site.

Fig. 7.7 shows how the Control worksheet is slightly modified from that shown earlier in Fig. 6.14 so that simulations is now for ten daily time steps, starting from Sept. 10, 2000 (cells B2 and B5:B8).

	A	B	C	D	E	F
1	INPUT					
2	date	=DATE(2000,9,10)		current date	=B2+_step	
3				doy	=date-DATE(YEAR(date),1,0)	
4	CONTROL			hour		
5	maxsteps	10				
6	stepsize	1				
7	step					
8	criteria	= _step<B5				
9						

Fig. 7.7. Control worksheet: slight modifications in the Control worksheet for daily simulations for ten days.

Note that cell E2 (date; Table 6.2) is modified to contain the formula: "=B2+_step" so that the simulation date can be incremented by one day (note: cell B6 or _stepsize is set to 1).

Cell E4 (th; Table 6.2) is left blank because this cell is used by the ITG action when it integrates Eq. 7.20 (cell Photosynthesis! J19) over the time period of sunrise to sunset.

Recall that cells A3 and A8 in the Output worksheet are defined with the cell names _read and _write, respectively (Table 6.2). However, the parameters to be included in the output listing have been modified to include the air temperature, solar irradiance parameters, and most importantly, the gross canopy assimilation rate (Fig. 7.8).

	A	B	C	D	E	F	G	H	I
1	TO OUTPUT								
2	doy	T_{min}	T_{max}	T_{mean}	$I_{df,d}$	$I_{dr,d}$	$I_{t,d}$	$A_{canopy,d}$	
3	=doy	=Tmin	=Tmax	=Tmean	=Idfd/10^6	=Idrd/10^6	=Itd/10^6	=dayassim	
4									
5									
6									
7	OUTPUT								
8									
9									
10									

Fig. 7.8. Output worksheet: setup of the what-to-output and output sections.

The division by 10^6 in cells E3, F3, and G3 is to convert the daily diffuse, direct, and total solar irradiance to MJ m^{-2} day^{-1}, respectively.

Once model implementation is completed, choose "Start Simulation" from the BuildIt menu to start the model simulations.

The model results (Fig. 7.9) show that the minimum and maximum air temperatures are rather constant at about 22 and 33 °C, respectively. However, total daily solar irradiance varies between about 10 to 20 MJ m^{-2} ground day^{-1}.

	A	B	C	D	E	F	G	H
1	TO OUTPUT							
2	doy	T_{min}	T_{max}	T_{mean}	$I_{df,d}$	$I_{dr,d}$	$I_{t,d}$	$\Lambda_{canopy,d}$
3	264	22.2	31.6	26.9	10.272	11.718	21.99	38.640997
4								
5								
6								
7	OUTPUT							
8	254	22.3	32.1	27.2	11.188	6.9747	18.163	36.854272
9	255	22.1	33	27.55	10.951	3.1974	14.149	29.95886
10	256	22.6	33.7	28.15	11.238	4.9934	16.232	32.558541
11	257	23	33.5	28.25	10.364	1.9113	12.275	26.19998
12	258	23.2	33.2	28.2	10.48	2.0588	12.539	26.833268
13	259	24	29.2	26.6	9.5856	1.2007	10.786	25.69616
14	260	22.6	33.6	28.1	9.5871	1.2009	10.788	23.398387
15	261	23	32.9	27.95	11.25	5.0045	16.255	33.162719
16	262	23.1	33.2	28.15	11.049	8.2667	19.316	37.214726
17	263	22.6	33.8	28.2	10.702	10.048	20.751	38.640997
18								
19								
20								
21								
22								
23								
24								
25								
26								
27								
28								
29								
30								
31								

Fig. 7.9. Simulation results in the Output worksheet.

As total solar irradiance increases or decreases, the gross canopy photosynthesis would likewise increase or decrease, respectively. The chart drawn from the model output confirms the close relationship between these two properties.

7.4 Exercises

1. Run the model to determine the daily canopy photosynthesis for Serdang starting from 10 Sept. 2000 for ten days for the following conditions:

   a) Intercellular CO_2 concentration, C_i, increasing from 270 to 350 μmol mol^{-1}.

   b) Canopy temperature, T_f, increasing from (T_a+2) to (T_a+4) °C.

   c) Both (a) and (b) conditions.

   Examine the model outputs to determine how increasing CO_2 concentration levels and air temperature affect canopy photosynthesis.

2. Eq. 7.23 integrates the canopy photosynthesis equation from sunrise (t_{sr}) to sunset (t_{ss}) because photosynthesis would not occur before sunrise or after sunset hours. However, photosynthesis becomes significant only when the solar elevation, β, is 6° or more above the horizon.
   This is why some models perform the canopy photosynthesis integration, not from sunrise to sunset hours, but from the hour just after sunrise when the sun is 6° above the horizon ($t_{sr,6°}$) to the hour just before sunset when the sun is again 6° above the horizon ($t_{ss,6°}$). In other words, integration is

$$\int_{t_{sr,6°}}^{t_{ss,6°}} \Lambda_{canopy}\ dt$$

Consequently, modify the model to determine the two hours in a day when $\beta = 6°$ and perform the canopy integration over the period between these two hours to determine the daily canopy photosynthesis for Serdang starting from 10 Sept. 2000 for ten days.

Compare the results with that when canopy integration was performed over the period between sunrise to sunset hours.

3. Modify the model so that intercellular CO_2 concentration, C_i, is determined by the equation from Yin and van Laar (2005) as:

$$C_i = C_a \left[1 - \left(1 - \Gamma^*/C_a \right)\left(a + bD_L \right) \right]$$

where C_a is the ambient CO_2 concentration (μmol mol^{-1}); Γ^* is the CO_2 compensation point (μmol mol^{-1}) (Eq. 7.13); and D_L is the leaf vapor pressure deficit (mbar) which is determined by

$$D_L = e_s \left[\!\left[T_f \right]\!\right] - e_a$$

where $e_s \left[\!\left[T_f \right]\!\right]$ is the saturated vapor pressure (mbar) (Eq. 6.19) at foliage temperature T_f (°C); and e_a is the air vapor pressure (mbar) (Eq. 6.21).

Simulate the daily canopy photosynthesis for Serdang starting from 10 Sept. 2000 for ten days, using C_a as 389 μmol mol^{-1} and coefficients a and b as 0.14 and 0.0116, respectively.

4. From Question 3: further modify the model so that

4. From Question 3: further modify the model so that given a date, the ambient CO_2 concentration, C_a, can be calculated. Set C_a at 338 μmol mol^{-1} for the year 1980, and for every year since then, C_a increases by 1.7 μmol mol^{-1} per year.

   Simulate the daily canopy photosynthesis for Serdang starting from 10 Sept. 2000 for ten days, using the same values as before, including the coefficients a and b set at 0.14 and 0.0116, respectively.

Chapter 8. Energy balance component

The energy balance of the soil-plant-atmosphere system is described using an electrical network analogy, where heat fluxes traverse a series of resistances to reach a reference level above the canopies (Fig. 8.1).

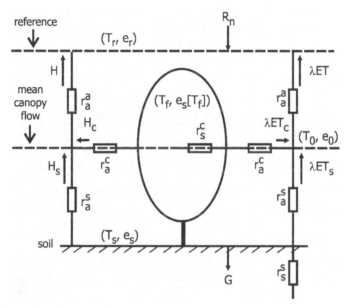

Fig. 8.1. Energy fluxes. Key: λET and H are the total latent heat and sensible heat fluxes, respectively (their subscripts 'c' and 's' indicate fluxes from the crop and soil, respectively); R_n is the net radiation flux into the system; G is the heat conduction into the soil; T_r, T_f, T_o, and T_s are the temperatures for the reference height, crop, mean canopy flow, and soil, respectively; e_r, e_s, and e_o are the vapor pressure at the reference height, soil surface, and mean canopy flow, respectively; $e_s[\![T]\!]$ is the saturated vapor pressure at temperature T; r_a^a and r_a^s are the aerodynamic resistance from the mean canopy flow to the reference height and from the soil to the mean canopy flow, respectively; r_a^c is the bulk boundary layer resistance; r_s^c and r_s^s are the canopy and soil surface resistance, respectively.

199

Equations in this chapter are taken (some adapted) from Kustas and Norman (1999), Farahani and Ahuja (1996), Goudriaan and van Laar (1994), Hansen (1993), Stannard (1993), Lafleur and Rouse (1990), Choudhury and Monteith (1988), Shuttleworth and Wallace (1985), Goudriaan (1977), Mitchell (1976), and Szeicz and Long (1969).

The various properties calculated by the energy balance component is visually depicted in Fig. 8.2.

8.1 Equations

8.1.1 Energy available to crop and soil

$$A = A_c + A_s$$
$$= R_n - G \tag{8.1a}$$

$$A_c = (1 - p_{Rn}) R_n \tag{8.1b}$$

$$A_s = p_{Rn} R_n - G \tag{8.1c}$$

$$p_{Rn} = \exp(-k_{Rn} L) \tag{8.1d}$$

where A is the net radiation available to the system (soil and crop); A_c is the fraction of net radiation available to the crop (W m^{-2} ground); A_s is the fraction of net radiation available to the soil (W m^{-2} ground); R_n is the net radiation (W m^{-2} ground) or Eq. 6.18; G is the soil heat flux (W m^{-2} ground); and p_{Rn} is the canopy penetration probability for net radiation which is dependent on L, the leaf area index (m^2 leaf m^{-2} ground), and k_{Rn}, the canopy extinction coefficient for net radiation (taken as 0.3).

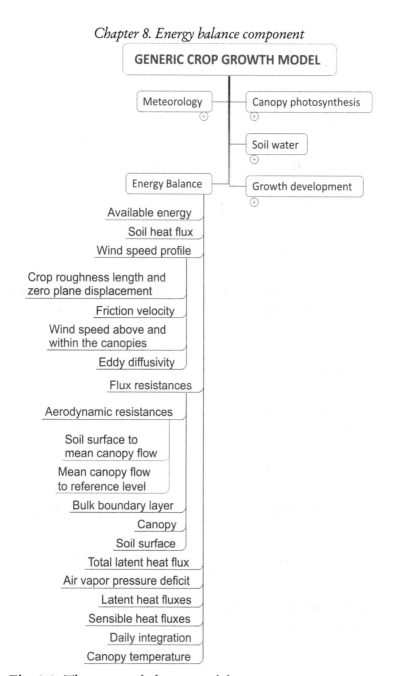

Fig. 8.2. The energy balance model component.

8.1.2   Soil/ground heat flux

$$G = 0.3 p_{Rn} R_n \qquad (8.2)$$

where G is the soil heat flux (W m^{-2} ground); p_{Rn} is the canopy penetration probability for net radiation (Eq. 8.1d); and R_n is the net radiation (W m^{-2} ground).

8.1.3   Wind speed profile

This section describes how wind speed varies with height above and within the plant canopies.

8.1.3.1 Crop roughness length and zero plane displacement

$$d = 0.64h \qquad (8.3)$$

$$z_0 = 0.26(h - d) \qquad (8.4)$$

where d is the zero plane displacement height (m); z_0 is the crop roughness length (m); and h is the plant height (m).

8.1.3.2 Friction velocity

$$u_* = ku[\![z_r]\!] \Big/ \ln\left(\frac{z_r - d}{z_0}\right) \qquad (8.5)$$

where u_* is the friction velocity (m s^{-1}); k is the von Karman constant (0.4); and $u[\![z_r]\!]$ is the wind speed (m s^{-1}) at the reference height z_r (m); d is the zero plane displacement height (m); z_0 is the crop roughness length (m).

8.1.3.3 Wind speed above and within the canopies

$$u[\![z]\!] = \begin{cases} \dfrac{u_*}{k}\ln\left(\dfrac{z-d}{z_0}\right) & z \geq h \\[2ex] \dfrac{u_*}{k}\ln\left(\dfrac{h-d}{z_0}\right)\exp\left[-n_u\left(1-\dfrac{z}{h}\right)\right] & z < h \end{cases} \tag{8.6}$$

where $u[\![z]\!]$ is the wind speed (m s^{-1}) at height z (m); h is the plant height (m); d is the zero plane displacement height (m); $z0$ is the crop roughness length (m); k is the von Karman constant (0.4); u_* is the friction velocity (m s^{-1}); and n_u is the wind speed extinction coefficient (taken as 2).

8.1.4   Flux resistances

The heat fluxes have to traverse various types of resistances within the soil-plant-atmosphere system. This section describes each of those resistance types.

8.1.4.1 Eddy diffusivity

$$K[\![z]\!] = K_h \cdot \exp\left[-n_K\left(1-\dfrac{z}{h}\right)\right] \tag{8.7}$$

$$K_h = ku_*h \tag{8.8}$$

where $K[\![z]\!]$ and Kh are the eddy diffusivities at height z and at the canopy top, respectively (m^2 s^{-1}); h is the plant height (m); k is the von Karman constant (0.4); u_* is the friction velocity (m s^{-1}); and n_K is the eddy diffusivity extinction coefficient (taken as 2).

8.1.4.2 Aerodynamic resistances

Soil surface to mean canopy flow

$$r_a^s = \frac{h \cdot \exp(n_K)}{n_K \cdot K_h}\left[\exp\left(-n_K \frac{z_{s0}}{h}\right) - \exp\left(-n_K \frac{z_0 + d}{h}\right)\right] \quad (8.9)$$

where r_a^s is the resistance between the soil surface and the mean canopy flow (s m^{-1}); h is the plant height (m); k is the von Karman constant (0.4); $u*$ is the friction velocity (m s^{-1}); n_K is the eddy diffusivity extinction coefficient (taken as 2); K_h is the eddy diffusivity at the canopy top (m^2 s^{-1}); d is the zero plane displacement height (m); z_0 is the crop roughness length (m); and z_{s0} is the soil surface roughness length (m).

Note: for flat, tilled land, z_{s0} can be taken as 0.004 m.

Mean canopy flow to reference level

$$
\begin{aligned}
r_a^a = &\frac{1}{ku_*}\ln\left(\frac{z_r - d}{h - d}\right) + \\
&\frac{1}{n_K ku_*}\left\{\exp\left[n_K\left(1 - \frac{z_0 + d}{h}\right)\right] - 1\right\}
\end{aligned}
\quad (8.10)
$$

where r_a^a is the resistance between the mean canopy flow and the reference level (s m^{-1}); h is the plant height (m); k is the von Karman constant (0.4); $u*$ is the friction velocity (m s^{-1}); n_K is the eddy diffusivity extinction coefficient (taken as 2); d is the zero plane displacement height (m); z_0 is the crop roughness length (m); and z_r is the reference height (m).

8.1.4.3 Bulk boundary layer resistance

$$r_a^c = \frac{n_u}{0.012L\left[1-\exp\left(-\dfrac{n_u}{2}\right)\right]\sqrt{\dfrac{u[\![h]\!]}{w}}} \tag{8.11}$$

where r_a^c is the bulk boundary layer resistance (s m^{-1}); L is the leaf area index (m^2 leaf m^{-2} ground); w is the mean leaf width (m); $u[\![h]\!]$ is the wind speed at the canopy top (*i.e.*, at plant height h) (m s^{-1}); and n_u is the wind speed extinction coefficient (taken as 2).

8.1.4.4 Canopy resistance

$$r_{st} = \frac{a_1 + 0.5I_t}{a_2\left(0.5I_t\right)} \tag{8.12}$$

$$r_s^c = \begin{cases} \dfrac{r_{st}}{L} & L \leq 0.5L_{max} \\[2ex] \dfrac{r_{st}}{0.5L_{max}} & L > 0.5L_{max} \end{cases} \tag{8.13}$$

where r_s^c is the canopy resistance (s m^{-1}); r_{st} is the leaf stomatal resistance (s m^{-1}); L is the leaf area index (m^2 leaf m^{-2} ground); L_{max} is the maximum total leaf area index (m^2 leaf m^{-2} ground); and I_t is the total hourly solar irradiance (W m^{-2} ground); a_1 and a_2 are empirical coefficients, dependent on the crop type.

Note: 50% of total solar irradiance is assumed to be PAR.

8.1.4.5 Soil surface resistance

$$r_{s,dry}^{s} = \frac{\tau l}{\phi_p D_{m,v}} \tag{8.14}$$

$$r_{s}^{s} = r_{s,dry}^{s} \exp\left(-\frac{1}{\lambda_p} \cdot \frac{\Theta_{v,1}}{\Theta_{v,sat,1}}\right) \tag{8.15}$$

where r_s^s is the soil surface resistance (s m^{-1}); $D_{m,v}$ is the vapor diffusion coefficient in air (24.7 x 10^{-6} m^2 s^{-1}); τ is soil tortuosity (taken as 2); ϕ_p is soil porosity; l is the dry soil layer thickness (taken as the first soil layer thickness) (m); $\Theta_{v,1}$ and $\Theta_{v,sat,1}$ are the volumetric soil water content and saturated soil water content (m^3 m^{-3}) of the first soil layer, respectively; and λ_p is the soil pore-size distribution index from the Brooks-Corey equation.

Note: the soil profile is divided into two or more layers (*see* next chapter).

8.1.5   Total latent heat flux

$$\lambda ET = C_c PM_c + C_s PM_s \tag{8.16a}$$

$$PM_c = \frac{\Delta A + \left(\rho c_p D - \Delta r_a^c A_s\right)/\left(r_a^a + r_a^c\right)}{\Delta + \gamma\left[1 + r_s^c/\left(r_a^a + r_a^c\right)\right]} \tag{8.16b}$$

$$PM_s = \frac{\Delta A + \left[\rho c_p D - \Delta r_a^s A_c\right]/\left(r_a^a + r_a^s\right)}{\Delta + \gamma\left[1 + r_s^s/\left(r_a^a + r_a^s\right)\right]} \tag{8.16c}$$

$$C_c = \left[1 + R_c R_a/R_s\left(R_c + R_a\right)\right]^{-1} \tag{8.16d}$$

$$C_s = \left[1 + R_s R_a/R_c\left(R_s + R_a\right)\right]^{-1} \tag{8.16e}$$

$$R_a = (\Delta + \gamma) r_a^a \tag{8.16f}$$

$$R_c = (\Delta + \gamma) r_a^c + \gamma r_s^c \tag{8.16g}$$

$$R_s = (\Delta + \gamma) r_a^s + \gamma r_s^s \tag{8.16h}$$

where λET is the total latent heat flux (W m^{-2} ground); r_a^a is the aerodynamic resistance between the mean canopy flow and reference height (s m^{-1}); r_a^s is the aerodynamic resistance between the soil and mean canopy flow (s m^{-1}); r_a^c is the bulk boundary layer resistance (s m^{-1}); r_s^c and r_s^s are the canopy and soil surface resistance, respectively (s m^{-1}); A, A_s and A_c are energy available to the system (total), soil and crop, respectively (W m^{-2} ground); Δ is the slope of the saturated vapor pressure curve (mbar K^{-1}); γ is the psychometric constant (0.658 mbar K^{-1}); D is the vapor pressure deficit (mbar); and ρc_p is the volumetric heat capacity for air (1221.09 J m^{-3} K^{-1}).

The slope of the saturated vapor pressure curve, Δ, is obtained from Eq. 6.20.

8.1.6 Air vapor pressure deficit at the mean canopy flow

$$D_0 = D + \frac{r_a^a}{\rho c_p} \left[\Delta A - (\Delta + \gamma) \lambda ET \right] \tag{8.17}$$

where D_0 is the vapor pressure deficit at the mean canopy flow (mbar); λET is the total latent heat flux (W m^{-2} ground); r_a^a is the aerodynamic resistance between the mean canopy flow and reference height (s m^{-1}); A is the total energy available to the system (W m^{-2} ground); Δ is the slope of the saturated vapor pressure curve (mbar K^{-1}); γ is the psychometric constant (0.658 mbar K^{-1}); D is the vapor pressure deficit (mbar); and ρc_p is the volumetric heat capacity for air (1221.09 J m^{-3} K^{-1}).

Knowing D_0 is essential because this value is used to calculate the latent and sensible heat fluxes for the soil and crop components.

8.1.7 Latent heat fluxes for crop and soil

$$\lambda ET_s = \frac{\Delta A_s + \rho c_p D_0 / r_a^s}{\Delta + \gamma \left(r_s^s + r_a^s\right)/r_a^s} \tag{8.18}$$

$$\lambda ET_c = \frac{\Delta A_c + \rho c_p D_0 / r_a^c}{\Delta + \gamma \left(r_s^c + r_a^c\right)/r_a^c} \tag{8.19}$$

where λET_s and λET_c are the soil and crop latent heat fluxes, respectively (W m^{-2} ground); A_s and A_c are energy available to the soil and crop, respectively (W m^{-2} ground); r_a^s is the aerodynamic resistance between the soil and mean canopy flow (s m^{-1}); r_a^c is the bulk boundary layer resistance (s m^{-1}); r_s^c and r_s^s are the canopy and soil surface resistance, respectively (s m^{-1}); Δ is the slope of the saturated vapor pressure curve (mbar K^{-1}); γ is the psychometric constant (0.658 mbar K^{-1}); D_0 is the vapor pressure deficit at the mean canopy flow (mbar); and ρc_p is the volumetric heat capacity for air (1221.09 J m^{-3} K^{-1}).

8.1.8 Sensible heat fluxes for crop and soil

$$H_s = \frac{\gamma A_s \left(r_s^s + r_a^s \right) - \rho c_p D_0}{\Delta r_a^s + \gamma \left(r_s^s + r_a^s \right)} \tag{8.20}$$

$$H_c = \frac{\gamma A_c \left(r_s^c + r_a^c \right) - \rho c_p D_0}{\Delta r_a^c + \gamma \left(r_s^c + r_a^c \right)} \tag{8.21}$$

where H_s and H_c are the soil and crop sensible heat fluxes, respectively (W m^{-2} ground); A_s and A_c are energy available to the soil and crop, respectively (W m^{-2} ground); r_a^s is the aerodynamic resistance between the soil and mean canopy flow (s m^{-1}); r_a^c is the bulk boundary layer resistance (s m^{-1}); r_s^c and r_s^s are the canopy and soil surface resistance, respectively (s m^{-1}); Δ is the slope of the saturated vapor pressure curve (mbar K^{-1}); γ is the psychometric constant (0.658 mbar K^{-1}); D_0 is the vapor pressure deficit at the mean canopy flow, respectively (mbar); and ρc_p is the volumetric heat capacity for air (1221.09 J m^{-3} K^{-1}).

8.1.8.1 Canopy temperature

Knowing the sensible heat fluxes would allow us to determine the canopy temperature as:

$$T_f = \frac{H_c r_a^c + \left(H_s + H_c \right) r_a^a}{\rho c_p} + T_r \tag{8.22}$$

where T_f is the canopy (foliage) temperature (°C); T_r is the air temperature at reference level (°C); H_s and H_c are the soil and crop sensible heat fluxes, respectively (W m^{-2}

ground); r_a^a is the aerodynamic resistance between the mean canopy flow and reference level (s m^{-1}); r_a^c is the bulk boundary layer resistance (s m^{-1}); and ρc_p is the volumetric heat capacity for air (1221.09 J m^{-3} K^{-1}).

8.1.9 Daily potential water loss

To determine the daily amount of water loss by soil evaporation and plant transpiration, we need to integrate Eq. 8.18 and 8.19, respectively, over 24 hours. In the five-point ($N=5$) Gaussian numerical integration method, five points within the 24-hour period in a day are selected, and for each selected hour, the latent heat fluxes are calculated.

However, most of the water losses occur during the day. Consequently, to avoid underestimation of the daily latent heat fluxes, we will integrate Eq. 8.18 and 8.19 over 12 hours instead and then multiply their respective results by 2. In other words, we will assume that the hourly latent heat fluxes are symmetric with respect at 12:00 hours.

So, to determine the amount of water loss by soil evaporation within a day, Eq. 8.18 is integrated as:

$$PET_s = \frac{2 \times 3600}{\lambda} \times \int_0^{12} \lambda ET_s dt \qquad (8.23)$$

where PET_s is the potential soil water loss by evaporation (mm day^{-1}); and λ is the latent heat of vaporization of water (2454000 J kg^{-1}). Integration is over the first half of the day, so multiplication by 2 is necessary to convert the answer into representing a 24-hour period. Multiplication by 3600 and division by λ is to convert the soil latent heat flux from having units W m^{-2} ground (equivalent to J m^{-2} ground s^{-1}) into mm day^{-1}.

Likewise, the daily water loss by plant transpiration is determined by integrating Eq. 8.19 as

$$PET_c = \frac{2 \times 3600}{\lambda} \times \int_0^{12} \lambda ET_c dt \qquad (8.24)$$

where PET_c is the potential water loss by plant transpiration (mm day^{-1}); and λ is the latent heat of vaporization of water (2454000 J kg^{-1}).

8.2 Measured parameters

The following are parameters that must be measured or supplied to the energy balance model component:

1. Reference height, z_r, the height of the weather station, is fixed at 2 m.
2. Soil roughness length, z_{s0}, is fixed at 0.004 m, which is the value for a freshly tilled soil.
3. The canopy extinction coefficient for net radiation (k_{Rn}) is taken as 0.3.
4. Values for the leaf stomatal resistance coefficients, a_1 and a_2, in Eq. 8.12 are 51.210 and 0.005, respectively.
5. Attenuation coefficients for wind speed (n_u) and eddy diffusivity (n_K) are both assumed to have a value of 2.
6. Soil properties: soil tortuosity (τ) is set at 2, soil porosity (ϕ_p) at 0.42 m^3 m^{-3}, and soil pore size distribution index (λ) at 0.18.
7. Provisional values:
   a) Because the crop growth development model component has yet to be built, we will set provisional values for plant height (h = 1 m),

211

leaf area index (as in section 7.2: $L = L_{max}$, where L_{max} is fixed at 3 m² m⁻²), and mean leaf width ($w = 0.08$ m).

b) The soil water model component too have yet to be built. Consequently, we will give provisional values for the current volumetric soil water content and saturated water content in the top soil layer as $\Theta_{v,1} = 0.2$ m³ m⁻³ and $\Theta_{v,sat,1} = 0.39$ m³ m⁻³, respectively, and the top soil dry layer thickness (l) as 0.02 m.

8.3 Implementation

In the gcg model workbook, insert a new worksheet and name it *ET* (Fig. 8.3).

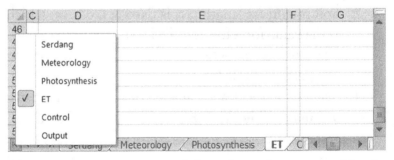

Fig. 8.3. Insert a new worksheet into the gcg workbook. This new worksheet is named *ET* and will contain equations pertaining to the energy balance of the system.

This brings the total number of worksheets in the gcg model workbook to six (Serdang, Meteorology, Photosynthesis, ET, Control, and Output worksheets).

We will implement Eq. 8.1 to 8.24 in the ET worksheet (Fig. 8.4 to 8.12). Again, to aid model readability, we will define and use several more cell names as listed in Table 8.1.

Table 8.1. Cell names defined for the energy balance model component, in addition to those already defined for the previous model components (Table 6.2 and 7.3).

Worksheet	Cell	Cell name
ET	B2	zr
	B3	zs0
	B4	kRn
	B5	stom_a1
	B6	stom_a2
	B7	nu
	B8	nK
	B16	psycho
	B17	k
	B18	pcp
	B19	Dmv
	E2	h
	E3	leafwidth
	E9	pRn
	E10	G
	E11	Ac
	E12	As
	E13	A
	E16	z0
	E17	d
	E18	ustar
	E19	uh
	E20	Kh
	E26	rsa
	E31	raa
	E35	rca

Worksheet	Cell	Cell name
	E38	rst
	E39	rcs
	E42	rss
	H13	LET
	H15	vpd0
	H19	LETs
	H21	PETs
	H25	LETc
	H27	PETc
	H29	PET
	H34	Hs
	H38	Hc

Fig. 8.4 shows the various model inputs and constants used in the energy balance model component.

Fig. 8.5 shows the calculations for the available energy for the soil and crop and the wind speed profile, based on Eq. 8.1 to 8.8. As discussed earlier, provisional values are given for some of the model parameters such as plant height, mean leaf width, and soil water content (cells E2:E6).

Fig. 8.6 shows the calculations for the various resistances based on Eq. 8.9 to 8.15. Note that cell E37 uses the MAX function to ensure no division by zero occurs when calculating the stomatal resistance in cell E38.

Fig. 8.7 and 8.8 show the calculations for the latent and sensible heat fluxes for the soil and crop.

Recall that we previously determined the canopy temperature (T_f) simply as 2 °C above the air temperature (T_a) (Fig. 7.4). Since we can now determine T_f more precisely (Fig. 8.9a), we should modify cell Photosynthesis!G3 to have the following formula:

"=ET!K2" so that Tf now refers to the more precise canopy temperature value (Fig. 8.9b). To minimize changes in the workbook, cell name Tf is unchanged and still refers to cell Photosynthesis!G3 (Table 7.3).

We will run daily simulations for 10 days beginning from Sept. 10, 2000 (Fig. 8.10). This part of the Control worksheet is unchanged from that in the previous chapter (*see* Fig. 7.7).

	A	B	C
1	INPUT		
2	z_r	2	
3	z_{s0}	0.004	
4	k_{Rn}	0.3	
5	stomata, a_1	51.21	
6	stomata, a_2	0.005	
7	n_u	2	
8	n_K	2	
9			
10	SOIL		
11	τ	2	
12	ϕ_p	0.42	
13	λ_p	0.18	
14			
15	CONSTANTS		
16	γ	0.658	
17	k	0.4	
18	ρc_p	1221.09	
19	$D_{m,v}$	=24.7*10^-6	

Fig. 8.4. ET worksheet: model inputs and constants.

	C	D	E	F
1		**PROVISIONAL**		
2		h	1	
3		w	0.08	
4		l	0.02	
5		$\Theta_{v,sat,1}$	0.39	
6		$\Theta_{v,1}$	0.2	
7				
8		**AVAIL. ENERGY**		
9		p_{Rn}	=EXP(-kRn*L)	
10		G	=0.3*pRn*Rn	
11		A_c	=(1-pRn)*Rn	
12		A_s	=pRn*Rn-G	
13		A	=Ac+As	
14				
15		**PROFILE**		
16		z_0	=0.13*h	
17		d	=0.64*h	
18		u*	=k*u/LN((zr-d)/z0)	
19		u[h]	=(ustar/k)*LN((h-d)/z0)	
20		K_h	=k*ustar*h	
21				

Fig. 8.5. ET worksheet: calculations for the available energy and wind speed profile.

	C	D	E	F
21				
22		**RESISTANCES**		
23		n1	=(h*EXP(nK))/(nK*Kh)	
24		n2	=EXP(-nK*zs0/h)	
25		n3	=EXP(-nK*(z0+d)/h)	
26		rsa	=E23*(E24-E25)	
27				
28		n4	=LN((zr-d)/(h-d))/(k*ustar)	
29		n5	=1/(nK*k*ustar)	
30		n6	=EXP(nK*(1-(z0+d)/h))-1	
31		raa	=E28+E29*E30	
32				
33		n7	=0.012*L*(1-EXP(-nu/2))	
34		n8	=SQRT(E19/leafwidth)	
35		rca	=nu/(E33*E34)	
36				
37		total PAR	=MAX(0.1,It*0.5)	
38		rst	=(stom_a1+E37)/(stom_a2*E37)	
39		rcs	=IF(L<=0.5*Lmax,rst/L,rst/(0.5*Lmax))	
40				
41		rss (dry)	=(B11*E4)/(B12*Dmv)	
42		rss	=E41*EXP(-(1/B13)*E6/E5)	
43				
44				
45				
46				

Fig. 8.6. ET worksheet: calculations of the various flux resistances.

	F	G	H	I
1		LATENT HEAT		
2		Ra	=(slopesvp+psycho)*raa	
3		Rc	=(slopesvp+psycho)*rca+psycho*rcs	
4		Rs	=(slopesvp+psycho)*rsa+psycho*rss	
5		C_c	=(1+(H3*H2)/(H4*(H3+H2)))^-1	
6		C_s	=(1+(H4*H2)/(H3*(H4+H2)))^-1	
7		n9	=slopesvp*A+(pcp*vpd-slopesvp*rca*As)/(raa+rca)	
8		n10	=slopesvp+psycho*(1+rcs/(raa+rca))	
9		PM_c	=H7/H8	
10		n11	=slopesvp*A+(pcp*vpd-slopesvp*rsa*Ac)/(raa+rsa)	
11		n12	=slopesvp+psycho*(1+rss/(raa+rsa))	
12		PM_s	=H10/H11	
13		λET	=H5*H9+H6*H12	
14				
15		D_0	=vpd+(raa/pcp)*(slopesvp*A-(slopesvp+psycho)*LET)	
16				
17		n13	=slopesvp*As+pcp*vpd0/rsa	
18		n14	=slopesvp+psycho*(rss+rsa)/rsa	
19		λET_s	=H17/H18	
20		$\int \lambda ET_s\, dt$		
21		E (mm day^{-1})	=H20*2*3600/2454000	
22				
23		n15	=slopesvp*Ac+pcp*vpd0/rca	
24		n16	=slopesvp+psycho*(rcs+rca)/rca	
25		λET_c	=H23/H24	
26		$\int \lambda ET_c\, dt$		
27		T (mm day^{-1})	=H26*2*3600/2454000	
28				
29		ET (mm day^{-1})	=PETs+PETc	

Fig. 8.7. ET worksheet: calculations for the latent heat fluxes.

218

	F	G	H	I
30				
31		SENSIBLE HEAT		
32		n17	=psycho*As*(rss+rsa)-pcp*vpd0	
33		n18	=slopesvp*rsa+psycho*(rss+rsa)	
34	H_s		=H32/H33	
35				
36		n19	=psycho*Ac*(rcs+rca)-pcp*vpd0	
37		n20	=slopesvp*rca+psycho*(rcs+rca)	
38	H_c		=H36/H37	
39				

Fig. 8.8. ET worksheet: calculations for the sensible heat fluxes.

a)

	I	J	K
1		CANOPY TEMP.	
2		T_f	=((Hc*rca+(Hs+Hc)*raa)/pcp)+Ta
3			

b)

	E	F	G
1			
2		*PROVISIONAL*	Change from "=Ta+2"
3		T_f	=ET!K2
4		L	=Lmax
5			

Fig. 8.9. Canopy temperature *Tf*. a) ET worksheet: calculations for a more precise canopy temperature, and b) Photosynthesis worksheet: change cell Photosynthesis!G3 from the provisional "=Ta+2" (*see* Fig. 7.4) to "=ET!K2" so that it now refers to the more precise determination of canopy temperature. Note: cell name Tf still refers to cell Photosynthesis!G3 (Table 7.3).

	A	B	C	D	E
1	INPUT				
2	date	=DATE(2000,9,10)		current date	=B2+_step
3				doy	=date-DATE(YEAR(date),1,0)
4	CONTROL			hour	
5	maxsteps	10			
6	stepsize	1			
7	step				
8	criteria	=_step<B5			
9					

Fig. 8.10. Control worksheet: simulations for 10 days beginning Sept. 10, 2000. This part of the worksheet is unchanged from that in Fig. 7.7.

	F	G	H	I	J	K	L	M	N	O
1		OPERATION								
2		ITG	=th	=assim	=Photosynthesis!J20	=tsr	=tss			
3		ITG	=th	=LETs	=ET!H20	0	12			
4		ITG	=th	=LETc	=ET!H26	0	12			
5										

Fig. 8.11. Control worksheet: an additional two more integrations (using ITG action) to determine the daily water loss from soil evaporation (cells G3:N3) and plant transpiration (cells G4:N4). Cell G2 is defined with the cell name _operation.

	A	B	C	D	E	F	G	H
1	TO OUTPUT							
2	doy	$\Lambda_{canopy,d}$	E	T	ET			
3	=doy	=dayassim	=PETs	=PETc	=PET			
4								
5								
6								
7	OUTPUT							
8								
9								
10								

Fig. 8.12. Output worksheet: the day of year, daily photosynthetic rate, and water losses (cells A3:E3) are to be included in the model output. Cell A3 and A8 are defined with cell names _read and _write respectively.

To solve Eq. 8.23 and 8.24, we require an additional two ITG actions in the operation section (Fig. 8.11). Recall cell G2 is defined with the cell name _operation (Table 7.3), so this cell marks the start of the operation section. The first ITG action (cells G2:N2) in the operation section, as discussed previously (*see* Fig. 7.6), is to determine the daily canopy photosynthesis.

The second ITG action is to determine the daily soil evaporation. Cells G3:N3 specify that Eq. 8.18 is to be integrated, with respect to the hour (th), from 0:00 to 12:00 hours, and the integration result is to be stored in cell ET!H20. Cell M3 is blank, so it is TRUE by default, and a single integration is performed. Cell N3 is also blank, so this second ITG action is always performed. Fig. 8.7 shows that cell H21 completes the calculations in Eq. 8.23 by multiplying the integration result by $(2 \times 3600 / \lambda)$ to determine the daily water loss from the soil.

Similarly, the third ITG action (cells G4:N4) tells BuildIt to integrate Eq. 8.19 to determine the daily plant transpiration, and the result is to be stored in cell ET!H26, after which this value will be multiplied by $(2 \times 3600 / \lambda)$ in cell ET!H27 (Fig. 8.7) to complete the calculations in Eq. 8.24.

Fig. 8.12 shows that we wish to include the day of year, daily canopy photosynthesis, water loss from the soil and canopy, and the total water loss from the system (cells A3:E3), and the model output will appear starting in cell A8. Recall that cells A3 and A8 are defined with the cell names _read and _write, respectively (Table 6.2).

Once model implementation is complete, click the "Start Simulation" command from the BuildIt menu. Simulation will run in daily time steps, starting from Sept.

10, 2000, for ten days. The model output results are as shown in Fig. 8.13.

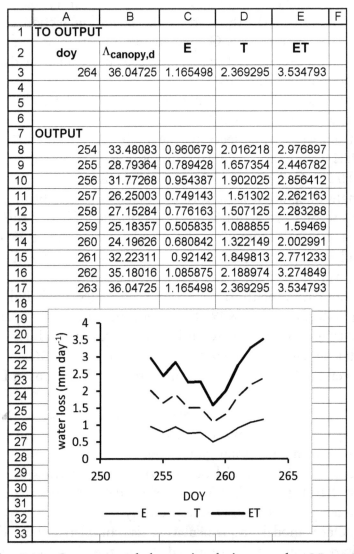

	A	B	C	D	E	F
1	TO OUTPUT					
2	doy	$\Lambda_{canopy,d}$	**E**	**T**	**ET**	
3	264	36.04725	1.165498	2.369295	3.534793	
4						
5						
6						
7	OUTPUT					
8	254	33.48083	0.960679	2.016218	2.976897	
9	255	28.79364	0.789428	1.657354	2.446782	
10	256	31.77268	0.954387	1.902025	2.856412	
11	257	26.25003	0.749143	1.51302	2.262163	
12	258	27.15284	0.776163	1.507125	2.283288	
13	259	25.18357	0.505835	1.088855	1.59469	
14	260	24.19626	0.680842	1.322149	2.002991	
15	261	32.22311	0.92142	1.849813	2.771233	
16	262	35.18016	1.085875	2.188974	3.274849	
17	263	36.04725	1.165498	2.369295	3.534793	
18						

Fig. 8.13. Output worksheet: simulation results. Note: E, T, and ET denote water loss from the soil, crop, and the system (soil and crop), respectively.

Column B shows the values for the daily canopy photosynthesis which are slightly different from that shown previously (Fig. 7.9). This is because canopy temperature (T_f) is now determined more precisely based on the sensible heat fluxes (Eq. 8.22) unlike before when we merely took T_f as always 2 °C higher than the current air temperature (T_a).

Columns C and D show the daily water loss from the soil and crop, respectively. Column E is the total water loss from both soil and crop. Water loss from the plant (T) is about double that from the soil (E). This is because of the high leaf area index (3 m^2 m^{-2}) which means near or full canopy cover. Within the ten-day period, total daily water loss from both soil and crop (ET) varied between about 1.5 to 3.5 mm day^{-1}.

8.4   Exercises

1. Modify the model so that the daily sensible fluxes for crop and soil can be determined.
   Hint: integration of the sensible fluxes over the whole day is required.

2. Run the model to simulate the *hourly* values for the energy available to the crop and soil (A_c and A_s), latent and sensible fluxes for the crop and soil (λET_c, λET_s, H_c, and H_s), air and canopy temperatures (T_a and T_f), the various flux resistances (r_a^a, r_a^s, r_a^c, r_s^c, and r_s^s), vapor pressure deficits (D and D_0), and net radiation (R_n). Simulation is for Serdang on 10 Sept. 2000 from 0.0 to 24.0 hours at every 1-hour interval.

3. From Question 2, repeat the simulations but for $L = 1$ m^2 leaf m^{-2} ground. What are the notable differences between the model outputs for $L = 1$ and 3 m^2 leaf m^{-2} ground, and why?

4. Sensitivity analysis can be done by increasing or decreasing the value of a parameter by 50 or 100% and determining how much this change affects the model output.

   Let us take, for instance, two variables, x_1 and x_2. When x_1 is increased by 50%, this change causes a 20% change in the model output, whereas a 50% change in x_2 only causes a 5% change in the model output. Consequently, the model output is considered to be more sensitive to x_1 than x_2.

   Consequently, modify the model so that you can determine to which of the five flux resistances, r_a^a, r_a^s, r_a^c, r_s^c, and r_s^s, the total latent heat, λET, is most (as well as least) sensitive.

Chapter 9. Soil water component

9.1 Multiple soil layers

For a more accurate simulation of the soil water content, the soil profile ought to be divided into two or more layers, so that the soil water content in each soil layer is determined (Fig. 9.1).

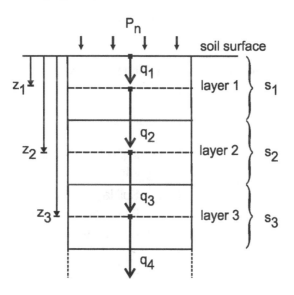

s_i = thickness of soil layer i (m)
z_i = depth of the middle of soil layer i from the surface (m)
q_i = water flux into soil layer i (mm day^{-1})
P_n = net rainfall (mm day^{-1})

Fig. 9.1. A soil profile is divided into three layers, and the water content in each layer is determined.

Simulation of soil water profile is intensive and time-consuming because it involves many repetitious calculations over a small time interval, mainly because soil hydraulic conductivity is sensitive to changes in the soil water content.

Equations in this chapter are taken (some adapted) from Teh (2006), Miyazaki (2005), Campbell (1994), Kropff (1993), and van Keulen and Seligman (1987).

The various properties calculated by the soil water model component is visually depicted in Fig. 9.2.

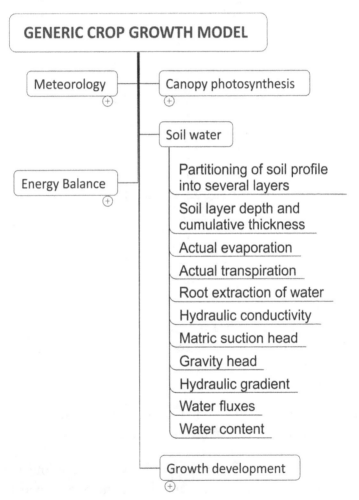

Fig. 9.2. Soil water model component.

9.2 Equations

9.2.1 Partitioning of soil profile into several layers

Fig. 9.1 shows that the soil profile is divided from the soil surface into three (N = 3) consecutive layers, where soil layer i (i = 1 to 3) has a thickness of s_i (m), and the depth from the soil surface to the middle of layer i is z_i (m). Water flux into soil layer i is denoted as q_i (mm day^{-1}).

The first soil layer is made thin (20 mm or 0.02 m) because this is the layer of soil that is in direct contact or interaction with the atmosphere, and soil evaporation is assumed to occur entirely out of this first soil layer. The entire soil profile is 1 m deep.

The following are assumptions we will use to determine the soil water content:

1. All soil layers have equal soil properties with one another, such as having the same soil texture and water retention characteristics.
2. No water table effect, which means that there is no capillary rise of water from the groundwater into the soil profile.
3. No surface runoff occurs.
4. The soil below the last soil layer is uniformly wet, and it has the same water content as the last soil layer. This means that water flows out of the last soil layer (*i.e.*, q_4) due to gravitational pull alone (*i.e.*, no matric suction gradient).

9.2.2 Soil layer depth and cumulative thickness

$$z_i = \sum_{j=1}^{i} 0.5\left(s_j + s_{j-1}\right) \quad \text{where } s_0 = 0 \tag{9.1}$$

where z_i is the depth of soil layer i from the soil surface (m), taken from the middle of that soil layer to the soil surface; s_j and s_{j-1} are the thickness of soil layer j and $j-1$, respectively (m).

9.2.3 Cumulative thickness

$$S_i = \sum_{j=1}^{i} s_j \tag{9.2}$$

where S_i is the cumulative thickness of soil layer i (m); that is, the sum of the thickness of layer i and all its preceding layers.

9.2.4 Actual evaporation

$$AET_s = PET_s \cdot R_{D,E} \tag{9.3a}$$

$$R_{D,E} = \frac{1}{1 + \left(3.6073 \dfrac{\Theta_{v,1}}{\Theta_{v,sat,1}}\right)^{-9.3172}} \tag{9.3b}$$

where AET_s is the actual evaporation (mm day^{-1}); PET_s is the potential evaporation (mm day^{-1}) (Eq. 8.23); $R_{D,E}$ is the reduction factor for evaporation (0 to 1); and $\Theta_{v,1}$ and $\Theta_{v,sat,1}$ are the current soil water content in the first soil layer ($i=1$) and saturated soil water content, respectively (m^3 m^{-3}).

Note: the soil water content from the first soil layer ($i=1$), $\Theta_{v,1}$, is taken because all evaporation occurs from this layer only (as this first layer is in contact with the atmosphere).

228

9.2.5   Actual transpiration

$$AET_c = PET_c \cdot R_{D,T} \tag{9.4a}$$

$$R_{D,T} = \begin{cases} 1 & \Theta_{v,root} \geq \Theta_{v,cr} \\ \dfrac{\left(\Theta_{v,root} - \Theta_{v,wp}\right)}{\left(\Theta_{v,cr} - \Theta_{v,wp}\right)} & \Theta_{v,wp} < \Theta_{v,root} < \Theta_{v,cr} \\ 0 & \Theta_{v,root} \leq \Theta_{v,wp} \end{cases} \tag{9.4b}$$

$$\Theta_{v,cr} = \Theta_{v,wp} + \frac{1}{2}\left(\Theta_{v,sat} - \Theta_{v,wp}\right) \tag{9.4c}$$

where AET_c is the actual transpiration (mm day^{-1}); PET_c is the potential transpiration (mm day^{-1}) (Eq. 8.24); $R_{D,T}$ is the reduction factor for transpiration (0 to 1); $\Theta_{v,root}$ is the total volumetric water content in the root zone (m^3 m^{-3}); $\Theta_{v,wp}$ is the volumetric water content at permanent wilting point (m^3 m^{-3}); and $\Theta_{v,sat}$ is the volumetric water content at saturation (m^3 m^{-3}).

The algorithm to determine the total volumetric water content in the root zone, $\Theta_{v,root}$, is as follows:

$$\Theta_{v,root} = \frac{1}{d_{root}} \times \sum_{i=1}^{N} MAX\left[0,\ \Theta_{v,i}\left(s_i - n_i\right)\right] \tag{9.4d}$$

$$n_i = MAX\left(0,\ S_i - d_{root}\right) \tag{9.4e}$$

where d_{root} is the rooting depth (m); s_i is the thickness of soil layer i (m); S_i is the cumulative thickness of soil layer i (m); and MAX function is the maximum of the enclosed values.

9.2.6 Extraction of water by roots

The algorithm to determine the amount of water extracted by roots in each soil layer is as follows:

$$AET_{c,i} = \left(\varphi_i - \varphi_{i-1} \right) AET_c \quad \text{where } \varphi_0 = 0 \tag{9.5a}$$

$$\varphi_j = 1.8 c_j - 0.8 c_j^{\;2} \tag{9.5b}$$

$$c_j = MIN \left(1, \; \frac{S_j}{d_{root}} \right) \tag{9.5c}$$

where $AET_{c,i}$ and AET_c are the amount of water extracted by roots in soil layer i and the actual transpiration, respectively (mm day^{-1}); d_{root} is the rooting depth (m); S_j is the cumulative thickness of soil layer j (m); and MIN function is the minimum of the enclosed values.

9.2.7 Matric suction head

The soil matric suction in soil layer i ($H_{m,i}$, m) is read from the table depicting the relationship between volumetric water content and matric suction (Table 9.1a). Like before, no effect of soil hysteresis is assumed.

Note: at soil saturation, suction is taken at 0 m.

9.2.8 Hydraulic conductivity

The hydraulic conductivity in soil layer i (K_i, mm day^{-1}) is read from the table depicting the relationship between volumetric water content and hydraulic conductivity (Table 9.1b). No effect of soil hysteresis is assumed.

Table 9.1. Relationship used in this book between the volumetric soil water content (Θv) with a) soil matric suction (Hm), and b) hydraulic conductivity (K).

a)

Θv (m^3 m^{-3})	Hm (m)
0.01	1250
0.12	150
0.13	50
0.17	10
0.19	4
0.20	3
0.23	2
0.28	1
0.29	0.5
0.39	0

b)

Θv (m^3 m^{-3})	K (mm day^{-1})
0.01	4.75×10^{-28}
0.05	3.90×10^{-14}
0.10	9.84×10^{-9}
0.15	1.42×10^{-5}
0.20	2.48×10^{-3}
0.25	0.14
0.30	3.58
0.35	56.92
0.39	396.79

9.2.9   Gravity head

$$H_{g,i} = z_i \qquad (9.6)$$

231

where $H_{g,i}$ is the gravity head in soil layer i (m); and z_i is the depth of the middle of soil layer i from the soil surface (m).

9.2.10   Hydraulic gradient

$$H_i = H_{m,i} + H_{g,i} \qquad (9.7)$$

where in soil layer i, H_i is the hydraulic gradient (m), $H_{m,i}$ is matric suction head (m), and $H_{g,i}$ is the gravity head (m).

9.2.11   Water fluxes

Darcy's law is used to describe the water flow in the soil. Water flow is taken to occur from the middle of layer i-1 to the middle of layer i (Fig. 9.1) Water flux into soil layer i is thus:

$$q_i = \begin{cases} P_n - E_a - T_{a,i} & i = 1 \\ \overline{K}\dfrac{(H_i - H_{i-1})}{(z_i - z_{i-1})} - T_{a,i} & 1 < i \le N \\ K_N & i = N + 1 \end{cases} \qquad (9.8)$$

$$\overline{K}_i = \frac{(K_{i-1}s_{i-1}) + (K_i s_i)}{s_{i-1} + s_i} \qquad (9.9)$$

where subscripts i-1 and i denote soil layer i-1 and i, respectively; q_i is the water flux into soil layer i (mm day^{-1}); P_n is the net rainfall (mm day^{-1}); E_a is the actual daily soil evaporation (occurs only from the first soil layer) (mm day^{-1}); $T_{a,i}$ is the daily extraction of water by roots (actual plant transpiration) from soil layer i (mm day^{-1}); $\overline{K}_i$ is the weighted average of the hydraulic conductivities of layer i and i-1 (mm day^{-1}); K_i and s_i are the soil layer's hydraulic

232

conductivity (mm day^{-1}) and thickness (m) of soil layer i, respectively; and Hi is the hydraulic gradient of soil layer i (m).

Not all of the rain above the canopies (*i.e.*, gross rainfall) reaches the ground due to canopy interception of the rain. The amount of gross rainfall reaching the ground is called the net rainfall, and it is simply determined by

$$P_n = \left[1 - MIN\left(m, \frac{L}{L_{max}} m \right) \right] P_g \qquad (9.10)$$

where P_n and P_g are the net and gross rainfall, respectively (mm day^{-1}); L and L_{max} are the crop's current leaf area index and maximum leaf area index, respectively (m^2 leaf m^{-2} ground). The MIN function is used to ensure that the fraction of gross rainfall intercepted by the canopies do not exceed m.

In Eq. 9.8, positive values for qi denote downward water flow, whereas negative values upward flow (*i.e.*, flowing against gravity pull).

The water flux out of the last soil layer ($i = N$) is denoted by q_{N+1}, and from Eq. 9.8, q_{N+1} is merely equal to K_N because of the assumption that the soil below the last layer is uniformly wet and it has the same water content as the last soil layer (section 9.2.1).

Consequently, water flux is only due to gravity gradient (no matric suction gradient). In this case, $q_{N+1} = k_N$.

Net flux

$$\hat{q}_i = q_i - q_{i+1} \qquad (9.11)$$

where $\hat{q}_i$ is the net flux in soil layer i (mm day^{-1}); and q_i and q_{i+1} are the water flux into soil layer i and $i+1$, respectively (mm day^{-1}). A positive net flux means the net flow is downward, whereas a negative net flux denotes a net upward flow.

9.2.12 Soil water content

Soil water content in soil layer i is determined using Euler's method (*see* Eq. 4.1) as:

$$\Theta_{i,t+\Delta t} = \Theta_{i,t} + \hat{q}_i \Delta t \qquad (9.12)$$

where $\Theta_{i,t}$ and $\Theta_{i,t+\Delta t}$ are the water content (mm) in soil layer i at time step t and $t+\Delta t$, respectively; $\hat{q}_i$ is the net flux in soil layer i (mm day^{-1}); and Δt is the time interval.

Note that soil hydraulic conductivity is sensitive to soil water content (*e.g.*, *see* Table 9.1b). Consequently, it is vital that the water fluxes are calculated using very small time steps because using too large a time step (*e.g.*, $\Delta t = 1$ day) may cause unrealistically large and erratic fluctuations or changes in the soil water content, where the soil water content fluctuates between very low to very high values in just a short period.

To overcome this problem, the single time interval is divided into a large number of subintervals (*e.g.*, have 100 or 200 subintervals within Δt), so that the soil water content is simulated at each successive subinterval (*e.g.*, $\frac{\Delta t}{100}$ or $\frac{\Delta t}{200}$).

Choosing too many subintervals would cause excessive calculations (with little gains in accuracy) and also cause very slow model runs.

In the other extreme, having too few subintervals would instead cause, as mentioned earlier, large and erratic fluctuations in the soil water content.

The optimal number of subintervals to have depends largely on how sensitive the soil's hydraulic conductivity is to the soil's water content. Some trial-and-error is required to determine the optimal number of subintervals. See section 4.1 for a discussion about this issue.

Finally, to convert volumetric soil water content ($\Theta_{v,i}$ in m^3 m^{-3}) to height of water (Θ_i in mm) is as follows:

$$\Theta_i = \Theta_{v,i} \times s_i \times 1000 \tag{9.13}$$

where s_i is the thickness (m) of soil layer i, and multiplication by 1000 is to convert the unit from m to mm.

9.3 Measured parameters

The following are parameters that must be measured or supplied to the soil water model component:

1. The whole soil profile is 1 m deep. We will further divide the soil profile into three layers ($N=3$) with the first, second, and third layers having thicknesses (s_i) of 0.02, 0.48, and 0.50 m, respectively.
2. The whole soil profile is assumed to be homogenous. Thus, each soil layer in the profile has equal soil properties with one another in terms of
   a) the relationship between volumetric soil water content (Θ_v) with matric suction head (H_m) as given in Table 9.1a and with hydraulic conductivity (K) in Table 9.1b, and
   b) the initial volumetric soil water content fixed at 0.2 m^3 m^{-3} (the soil's field capacity).
3. The maximum fraction of gross rainfall intercepted by the canopies and do not reach the soil surface is $m=0.1$ (10%).

4. *Provisional value*: the rooting depth (d_{root}) is kept constant at 0.3 m.

   When we implement the growth model component (see next chapter), this rooting depth will increase with time as the crop grows.

9.4 Implementation

Insert two new worksheets in the gcg model workbook, and name the first and second new worksheet as *Tables* and *Water*, respectively (Fig. 9.3).

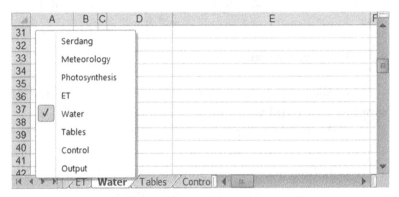

Fig. 9.3. Insert two new worksheets (and name them *Tables* and *Water*) in the gcg model workbook.

This brings the total number of worksheets in the gcg model workbook to eight. The Tables worksheet contains tabulated data (Fig. 9.4). For the moment, this worksheet contains two tables: the first table describes the relationship between volumetric soil water content with matric suction head (Table 9.1a) and the second between volumetric soil water content with hydraulic conductivity (Table 9.1b).

	A	B	C	D	E	F
1						
2	SOIL SUCTION			HYDRAULIC COND.		
3	$\Theta_{v,i}$	H_m (m)		$\Theta_{v,i}$	K (mm day^{-1})	
4	0.01	1250		0.01	4.75E-28	
5	0.12	150		0.05	3.90E-14	
6	0.13	50		0.1	9.84E-09	
7	0.17	10		0.15	0.0000142	
8	0.19	4		0.2	0.00248	
9	0.2	3		0.25	0.14	
10	0.23	2		0.3	3.58	
11	0.28	1		0.35	56.92	
12	0.29	0.5		0.39	396.79	
13	0.39	0				
14						

Fig. 9.4. Tables worksheet: create two tables to show the relationship between volumetric soil water content (m^3 m^{-3}) with matric suction head (m) (Table 9.1a) and with hydraulic conductivity (mm day^{-1}) (Table 9.1b).

Define the cell range A4:A13 and B4:B13 with cell names matric_vwc and matric_Hm, respectively (Table 9.2).

Likewise, define the cell range D4:D12 and E4:E12 with cell names hydraulic_vwc and hydraulic_K, respectively. These cell names will make referencing these two tables easier and improve model readability.

The second new worksheet, named Water, contains the calculations required to determine the soil water content in each of the three soil layers based on Eq. 9.1 to 9.13 (Fig. 9.5 to 9.10).

Fig. 9.5 shows the two model inputs for the soil water model component. Cell B2 is the initial volumetric soil water content and cell B3 the maximum fraction ($m = 0.1$) of gross rainfall that is intercepted by the canopies and that will not reach the ground. This *m* value is used in cell E17 to determine the net rainfall (Eq. 9.10). Until we have implemented the growth model component in the next chapter, cell E2 is the provisional value for the rooting depth (*d*root = 0.3 m). Cells E5:E6 calculate the actual evaporation from the soil (Eq. 9.3) and cells E9:E14 the actual plant transpiration (Eq. 9.4).

Table 9.2. Cell names defined for the soil water model component, in addition to those already defined for the previous model components (Table 6.2, 7.3, and 8.1).

Worksheet	Cell	Cell name
Control	G21	`_prerun`
Tables	A4:A13	`matric_vwc`
	B4:B13	`matric_Hm`
	D4:D12	`hydraulic_vwc`
	E4:E12	`hydraulic_K`
Water	B3	`m`
	E2	`droot`
	E6	`AETs`
	E14	`AETc`
	E17	`Pn`

Fig. 9.6 and 9.7 show the water flux calculations for each soil layer. Calculations are arranged such that column H contains the calculations for the first soil layer, J for the second, and K for the third. All calculations finally lead to the net water flux calculations (Eq. 9.11) in cells H22:J22 (Fig. 9.7).

	A	B	C	D	E	F
1	INPUT			*PROVISIONAL*		
2	Θ_V (initial)	0.2		d_{root}	0.3	
3	m	0.1				
4				ACTUAL E		
5				$R_{D,E}$	=1/(1+(3.6073*H9/H3)^-9.3172)	
6				AET_s	=PETs*E5	
7						
8				ACTUAL T		
9				mean $\Theta_{V,wp}$	=AVERAGE(H2:J2)	
10				mean $\Theta_{V,sat}$	=AVERAGE(H3:J3)	
11				$\Theta_{V,cr}$	=E9+0.5*(E10-E9)	
12				$\Theta_{V,root}$	=SUM(H16:J16)/droot	
13				$R_{D,T}$	=IF(E12>=E11,1,MAX(0,(E12-E9)/(E11-E9)))	
14				AET_c	=PETc*E13	
15						
16				NET RAIN		
17				P_n	=(1-MIN(m,L/Lmax*m))*Pg	
18						

Fig. 9.5. Water worksheet: model inputs and calculations for the actual water loss by soil evaporation and plant transpiration and for the net rainfall.

F	G	H	I	J	K
1	LAYER NO.	1	2	3	
2	$\Theta_{v,wp,i}$	=interpolate(matric_Hm,matric_vwc,150)	=interpolate(matric_Hm,matric_vwc,150)	=interpolate(matric_Hm,matric_vwc,150)	
3	$\Theta_{v,sat,i}$	=interpolate(matric_Hm,matric_vwc,0)	=interpolate(matric_Hm,matric_vwc,0)	=interpolate(matric_Hm,matric_vwc,0)	
4	s_i	0.02	0.48	0.5	
5	S_i	=H4	=I4+H5	=J4+I5	
6	z_i	=0.5*H4	=H6+0.5*(H4+I4)	=I6+0.5*(I4+J4)	
7	Θ_i (initial)	=$B$2*H4*1000	=$B$2*I4*1000	=$B$2*J4*1000	
8	Θ_i (current)				
9	$\Theta_{v,i}$ (current)	=MAX(0.01,MIN(H3,(H8/1000)/H4))	=MAX(0.01,MIN(I3,(I8/1000)/I4))	=MAX(0.01,MIN(J3,(J8/1000)/J4))	

Fig. 9.6. Water worksheet: the soil profile is divided into three layers, and for each soil layer, its water content is calculated.

240

F	G	H	I	J	K
1	LAYER NO. 1		2	3	
10	K_i	=interpolate(hydraulic_vwc,hydraulic_K,H9)	=interpolate(hydraulic_vwc,hydraulic_K,I9)	=interpolate(hydraulic_vwc,hydraulic_K,J9)	
11	$H_{m,i}$	=interpolate(matric_vwc,matric_Hm,H9)	=interpolate(matric_vwc,matric_Hm,I9)	=interpolate(matric_vwc,matric_Hm,J9)	
12	$H_{g,i}$	=H6	=I6	=J6	
13	H_i	=H11+H12	=I11+I12	=J11+J12	
14	mean K_i		=(H10*H4+10*I4)/(I4+H4)	=(I10*I4+J10*J4)/(J4+I4)	
15	n_i	=MAX(0,H5-droot)	=MAX(0,I5-droot)	=MAX(0,J5-droot)	
16	$\Theta_{v,i}(s_i-n_i)$	=MAX(0,H9*(H4-H15))	=MAX(0,I9*(I4-I15))	=MAX(0,J9*(J4-J15))	
17	$E_{a,i}$	=E6			
18	c_j	=MIN(1,H5/droot)	=MIN(1,I5/droot)	=MIN(1,J5/droot)	
19	φ_j	=1.8*H18-0.8*H18^2	=1.8*I18-0.8*I18^2	=1.8*J18-0.8*J18^2	
20	$AET_{c,i}$	=H19*AETc	=(I19-H19)*AETc	=(J19-I19)*AETc	
21	q_i	=Pn-H17-H20	=I14*(I13-H13)/(I6-H6)-I20	=J14*(J13-I13)/(J6-I6)-J20	
22	net q_i	=H21-I21	=I21-J21	=J21-J14	

Fig. 9.7. Water worksheet: calculations continue from Fig. 9.6, where all calculations finally lead to the determination of the net water fluxes (cells H22:J22). Note: rows 2 to 9 are hidden.

241

Once the net flux in each soil layer is calculated, the water content in each soil layer is updated using Eq. 9.12. Implementing Eq. 9.12 means the water content in soil layer 1 is updated by multiplying the value in cell H22 (net flux, $\hat{q}_1$) with Δt (_stepsize) and adding the result with the value in cell H8 (which holds the current soil water content, $\Theta_{1,t}$). The result is then stored in cell H8; thus, updating the soil water content to $\Theta_{1,t+\Delta t}$. The same calculations are performed for the other two soil layers.

To implement Eq. 9.12 in Excel, BuildIt's UPD action is required (*see* section 4.1 and Eq. 4.1). In the operation section in the Control worksheet, add the UPD action (Fig. 9.8).

	F	G	H	I	J	K	L	M	N	O
1		OPERATION								
2		ITG	=th	=assim	=Photosynthesis!J20	=tsr	=tss			
3		ITG	=th	=LETs	=ET!H20	0	12			
4		ITG	=th	=LETc	=ET!H26	0	12			
5		UPD	=Water!H8:J8	=Water!H22:J22	100					
6										
7										
19										
20		PRERUN								
21		INI	=Water!H8:J8	=Water!H7:J7						
22										
23										

Fig. 9.8. Control worksheet: the UPD action in the operation section is to determine the daily change in soil water content in each soil layer. The INI action in the prerun section is to initialize the soil water content at the start of a model run. Cells G2 and G21 are defined with cell names _operation (Table 7.3) and _prerun, respectively (Table 9.2). Note: rows 8 to 18 are hidden to show both the operation and prerun sections in a single screenshot.

Cells G5:K5 define the UPD action (Fig. 9.8). The UPD action will add each soil layer's current soil water content (the formula "=Water!H8:J8" in cell H5) with its net flux (the formula "=Water!H22:J22" in cell I5). Cell J5 reads 100 so that the soil water content will be calculated at every $\Delta t/100$ subinterval. We will set $\Delta t = 1$ day, so $\Delta t/100$ is then 14.4 mins, which means that the daily soil water content in each soil layer is updated (calculated) at every 14.4 mins. Depending on how sensitive hydraulic conductivity is to the soil water content, more than 100 subintervals may sometimes be required. But in this example, having 100 subintervals are sufficient.

As mentioned earlier, cells H8:J8 in the Water worksheet hold each soil layer's current soil water content. When the model is first run, we need to give each soil layer its initial soil water content. In this example, we set the initial soil water content in each layer to be 0.2 m^3 m^{-3} as specified in cell B2 (Fig. 9.5). This initial soil water content is then converted from volumetric soil water content (m^3 m^{-3}) to the height of water (mm) in cells H7:J7 using Eq. 9.13 (Fig. 9.6).

We now need a way to copy these initial values to cells H8:J8 when the model first runs. To perform this task, we will use BuildIt's INI action, and this action has to specified in the prerun section (*see* section 4.1).

In the Control worksheet, create the prerun section by defining the cell name _prerun to cell G21 (Fig. 9.8 and Table 9.2). Currently, only one INI action is specified (cells G21:J21) in this prerun section. This INI action copies the values from cells Water!H7:J7 (initial values) and pastes these values in cells Water!H8:J8 (current soil water content). Recall that any actions in the prerun section are executed only once by BuildIt.

243

In Fig. 9.6, the thickness for each soil layer (s_i) is specified in cell H4, I4, and J4. Cells H2:J2 and H3:J3 determine the volumetric soil water content at permanent wilting point (150 m suction) and at saturation point (0 m suction), respectively, by reading the table values specified in cells A4:B13 in the Tables worksheet (Fig. 9.4). BuildIt's `interpolate` function (*see* section 5.3.1) is used in the table lookup to ensure linear interpolation is performed if no exact match is found.

Note that because we are using cell names, we can improve model readability by using the named cell ranges in the table lookups. For example, in cell H10, the formula reads as: "`=interpolate(hydraulic_vwc, hydraulic_K, H9)`" which is easier to read than that without the use of named cell ranges: "`=interpolate(Tables!D4:D12, Tables!E4:E12, H9)`".

A slight modification is required in the ET worksheet (which we created in the previous chapter). Modify cells E4:E6 in the ET worksheet to that as shown in Fig. 9.9, so that these three cells now refer to the properties (thickness, soil water content at saturation point, and the current soil water content) of the first soil layer as specified in the Water worksheet. Previously, we gave these three soil properties provisional values (Fig. 8.5) because the soil water model component has yet to be built then.

In this example, we will simulate the daily (Δt or `_stepsize` = 1 day) soil water content for 365 days (1 year) starting from Jan. 1, 2000 (Fig. 9.10). Thus, in the Control worksheet, cell B2 is changed so that simulation starts at Jan. 1, 2000 and cell B5 is changed to 365 so that simulation runs for 365 days. Cell E2 is also modified to hold the formula "`=B2+INT(_step)`", where previously, the formula was "`=B2+_step`" (Fig. 7.7 and 8.10). This is a small but important modification, without which the model will give erroneous results.

	C	D	E	F
1		*PROVISIONAL*		
2		h	1	
3		w	0.08	
4		l	=Water!H4 ➝	
5		$\Theta_{v,sat}$	=Water!H3 ➝ No longer "0.02", "0.39", and "0.2", respectively	
6		$\Theta_{v,1}$	=Water!H9 ➝	
7				

Fig. 9.9. ET worksheet: with the implementation of the soil water model component, cells E4:E6 in the ET worksheet are changed so that they now refer to their respective values in the Water worksheet.

	A	B	C	D	E	F
1	INPUT					
2	date	=DATE(2000,1,1)		current date	=B2+INT(_step)	
3				doy	=date-DATE(YEAR(date),1,0)	
4	CONTROL			hour		
5	maxsteps	365				
6	stepsize	1				
7	step					
8	criteria	=_step<B5				
9						

Fig. 9.10. Control worksheet: simulation at daily time steps from Jan. 1, 2000 for 365 days (a year). Cell E2 is changed to use the INT function to ensure only integer values of _step are taken.

Excel's INT function is used in cell E2 (Fig. 9.10) to ensure that only integer values (*i.e.*, numbers without decimals) of _step are added to cell B2. Previously, we need not be concern of non-integer _step values because _stepsize (Δt) is unchanged at 1 (cell B6). Hence, _step will start as 0, then 1, 2, 3, and so on until the loop run terminates.

But our use of the UPD action, as discussed earlier, and having it divide Δt into 100 subintervals means _step can have non-integer values since _stepsize is not 1 but 0.01 (or 1/100). So, without the INT function, the formula "=B2+_step" will result in failed weather table lookups (Fig. 6.8) to obtain the meteorological properties for a given date.

The Output worksheet is also modified to include rain, the soil water content in each of the three soil layers, and the total water content in all the soil layers in the model output (Fig. 9.11). Note that the soil water content is to be outputted at the start (rather than at the end) of the current date; thus, the use of the FALSE values in cells C4:F4 (*see* section 4.1 and Fig. 4.2).

	A	B	C	D	E	F	G
1	TO OUTPUT						
2	doy	Rain	$\Theta_{v,1}$	$\Theta_{v,2}$	$\Theta_{v,3}$	total Θ	
3	=doy	=Pg	=Water!H9	=Water!I9	=Water!J9	=SUM(Water!H8:J8)	
4			FALSE	FALSE	FALSE	FALSE	
5							
6							
7	OUTPUT						
8							

Fig. 9.11. Output worksheet: model output to consist of day of year, gross rainfall, the volumetric soil water content for each of the three soil layers, and the total water content (in mm). Output listing will start at cell A8. Note: cell A3 is defined with cell name _read and cell A8 _write (Table 6.2).

Once the model implementation is complete, click the "Start Simulation" from the BuildIt menu, and the simulation results are as shown in Fig. 9.12.

	A	B	C	D	E	F	G
1	TO OUTPUT						
2	doy	Rain	$\Theta_{v,1}$	$\Theta_{v,2}$	$\Theta_{v,3}$	total Θ	
3	366	0	0.348999	0.355888	0.330389	343.001	
4			FALSE	FALSE	FALSE	FALSE	
5							
6							
7	OUTPUT						
8	1	11.8	0.2	0.2	0.2	200	
9	2	16	0.321635	0.214611	0.200011	209.452	
10	3	15.9	0.340549	0.240952	0.200164	222.55	
11	4	3.7	0.336049	0.264992	0.201618	234.726	
12	5	13	0.282665	0.265828	0.205695	236.098	
13	6	1	0.322495	0.279608	0.211082	246.203	
14	7	0	0.268866	0.273009	0.217052	244.947	
15	8	15.1	0.255029	0.264634	0.220474	242.362	
16	9	0	0.332524	0.281304	0.224327	253.84	
17	10	0	0.261593	0.272556	0.228912	250.515	
367	360	11	0.281719	0.28597	0.310808	298.304	
368	361	10.1	0.343865	0.291481	0.305514	299.545	
369	362	10	0.338319	0.296929	0.304217	301.401	
370	363	0	0.331163	0.300958	0.30364	302.903	
371	364	5	0.283823	0.288802	0.30276	295.681	
372	365	100.5	0.299508	0.28819	0.300306	294.474	
373							

Fig. 9.12. Output worksheet: simulation results of the daily water content for one year. Note: rows 18 to 366 are hidden to show the start and end portions of the model output.

With the implementation of the soil water component, simulation run has become noticeably longer to complete, not only because the simulation runs for 365 simulation days, but also because of the UPD action which we have instructed to update the soil water content at every 1/100 day (or every 14.4 mins). Having more soil layers or more frequent soil water content updates will slow down the simulation run even more.

From the model output, four charts are drawn to show how the soil water content in the three soil layers (as well as total soil water content) change with time and with every rainfall period (Fig. 9.13).

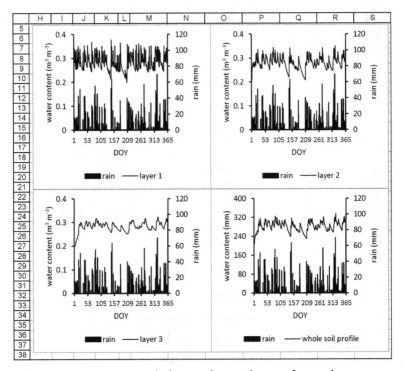

Fig. 9.13. Output worksheet: charts drawn from the output listing to show how the soil water content in the three soil layers change with time and rainfall. Note: DOY means day of year.

The charts in Fig. 9.13 show that Serdang experienced heavy rainfall throughout the year in 2000, with only short and infrequent periods of dry days (the longest dry period is no more than 14 days). The soil water content in all the three soil layers which we artificially set at an initial value of 0.2 m^3 m^{-3} quickly increased due to the large rainfall amount and frequent rainfall period. Soil water content in all three layers averaged at about 0.3 m^3 m^{-3} throughout the year. This level of soil water content is above the soil's water content at field capacity (0.2 m^3 m^{-3}). By the end of the year, the total soil water content increased from 200 mm to about 300 mm, a net annual increase of about 100 mm due to Serdang's heavy and frequent rains.

Soil water content would predictably increase after a rainfall period and decline during dry periods. The first soil layer showed the largest variation in soil water content since this soil layer is the uppermost layer and thus in direct contact with rainfall and the atmosphere (soil evaporation occurs solely from this soil layer). The deeper the layer in the soil, the soil water content is more stable and less responsive to changes in the weather, as we can see for the third and deepest soil layer in Fig. 9.13.

9.5 Exercises

1. Modify the model so that the rain is artificially set to zero (no rain throughout the simulation period). Run the model to determine how long the water content in each soil layer reaches permanent wilting point.
   Likewise, run the model to determine how long the whole soil profile reaches permanent wilting point. Recall: volumetric soil water content (m^3 m^{-3}) = mm water / soil depth (in mm).
   How would you modify the loop criteria in the Control worksheet so that the simulation ends when the water content in a given soil layer or whole soil profile reaches permanent wilting point?

2. Add a fourth layer in the soil profile. This soil layer has a thickness of 0.2 m and has the same properties as the other layers. Run the simulation as before.

3. Make the second soil layer a compacted layer with the following soil water retention and hydraulic conductivity properties:

Θ_v (m^3 m^{-3})	H_m (m)	K (mm day^{-1})
0.01	1250	3.0×10^{-28}
0.12	150	8.0×10^{-9}
0.25	3	0.1
0.30	0	3.0

Modify the model to have the second soil layer as a compacted layer with the above properties, and run the simulation as before to determine the soil water content in each soil layer. Note: the second soil layer is still 0.48 m as before.

What is the effect of the compacted second soil layer on the water content in each soil layer?

What is the implicit assumption in the model in regard to excess water (when the water content exceeds the soil's saturation point)?

4. Increase the rooting depth, d_{root}, from 0.3 to 0.9 m. Run the simulation as before.

What is the effect of a deeper rooting depth to the soil water content?

Chapter 10. Growth development component

Food produced via photosynthesis is used for plant maintenance (supporting processes for continual plant survival) and plant growth (synthesis of new cells). Equations in this chapter are taken (some adapted) from Goudriaan and van Laar (1994), Kropff (1993), and Spitters (1989). The various properties calculated by the crop growth development model component is visually depicted in Fig. 10.1.

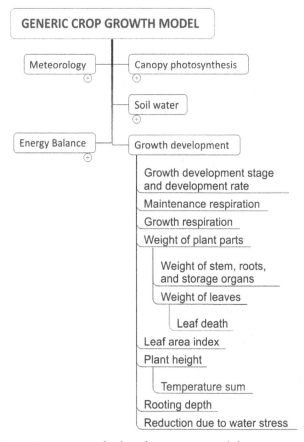

Fig. 10.1. Crop growth development model component.

10.1 Equations

10.1.1 Growth development stage and development rate

The growth of a crop can be distinguished into several important stages or "milestones" in its life cycle. These milestones can be the time of seed emergence, flowering, tuber initiation, bulking and maturity, ripening, plant maturity, and senescence.

Nonetheless, in this book, the crop growth stages are distinguished into just three points as shown in Table 10.1.

Table 10.1. The scale for the growth development stage, ξ_s, ranges from 0 to 2.

Scale	ξ_s
0	Planting
1	Flowering (anthesis)
2	Maturity

The growth stage of a crop progresses according to its growth development rate as

$$\xi_{s,t+\Delta t} = \xi_{s,t} + \xi_{r,t}\Delta t \tag{10.1}$$

where $\xi_{s,t}$ and $\xi_{s,t+\Delta t}$ are the growth development stage at time t and subsequent time step $t+\Delta t$; Δt is the interval for each time step (day); $\xi_{r,t}$ is the growth development rate (day^{-1}) at time t.

The growth development rate, ξ_r, is read from a table of values which describes the typical linear relationship between ξ_r with air temperature (Table 10.2). Note that in Table 10.2, no crop growth occurs ($\xi_r = 0$) at 0 °C and below. Consequently, the crop's base temperature T_b is 0 °C.

Table 10.2. Tabulated values of the crop growth development rate, ξ_r, according to air temperature for before ($\xi_s < 1$) and after ($\xi_s \geq 1$) anthesis. Note: T_b is the crop's base temperature, below which crop growth ceases.

Air temperature	ξ_r (day^{-1})	
(°C)	$\xi_s < 1$	$\xi_s \geq 1$
0 (T_b)	0.000	0.000
30	0.027	0.031
40	0.027	0.031

10.1.2 Maintenance respiration

$$R'_M = R_M \times 2^{(T_a - 25)/10} \times \frac{W_{GL}}{W_L} \qquad (10.2a)$$

$$R_M = \left(k_{GL}W_{GL}\right) + \left(k_S W_S\right) + \left(k_R W_R\right) + \left(k_O W_O\right) \qquad (10.2b)$$

where R'_M is the maintenance respiration requirement (g CH$_2$O m^2 ground day^{-1}); T_a is the air temperature (°C); k_i is the maintenance respiration coefficient (g CH$_2$O g^{-1} dry matter day^{-1}) for plant part i (Table 10.3); and W_i is the weight (g dry matter m^2 ground) for plant part i (where subscripts 'GL' , 'L', 'S', 'R', and 'O' denote green leaves, all leaves, stem, roots, and storage organs, respectively).

Maintenance respiration at 25 °C is calculated in Eq. 10.2b which is then is corrected for current air temperature and plant age in Eq. 10.2a.

Note that W_L is the total weight of leaves: weight of green (alive) leaves (W_{GL}) + weight of dead leaves (W_{DL}).

Table 10.3. Maintenance respiration coefficient (at 25 °C air temperature) for various plant parts.

Plant part i	k_i (g CH_2O g^{-1} dry matter day^{-1})
green leaves, k_{GL}	0.030
stem, k_S	0.015
roots, k_R	0.015
storage organs, k_O	0.010

10.1.3 Growth respiration

$$G_T = (F_{GL}G_{GL}) + (F_S G_S) + (F_R G_R) + (F_O G_O) \qquad (10.3)$$

where G_T is the total glucose requirement (g CH_2O g^{-1} dry matter); G_i is the glucose requirement (g CH_2O g^{-1} dry matter) for plant part i (Table 10.4); and F_i is the dry matter fraction of plant part i to the whole plant (Table 10.5).

Table 10.4. General glucose requirement for the synthesis of various plant parts.

Plant part i	G_i (g CH_2O g^{-1} dry matter)
green leaves, G_{GL}	1.463
stem, G_S	1.513
roots, G_R	1.444
storage organs, G_O	1.415

Table 10.5. Dry matter partitioning of various plant parts (F_i) according to the crop growth stage (ξ_s).

	Fraction of plant part i to whole plant			
	green leaves	stem	roots	storage organs
ξ_s	F_{GL}	F_S	F_R	F_O
0.0	0.76	0.15	0.09	0.00
0.2	0.67	0.21	0.12	0.00
0.4	0.66	0.29	0.05	0.00
0.6	0.62	0.34	0.04	0.00
0.8	0.38	0.53	0.09	0.00
1.0	0.27	0.51	0.17	0.05
2.0	0.00	0.00	0.00	1.00

10.1.4   Weight of plant parts (except leaves)

$$W_{i,t+\Delta t} = W_{i,t} + F_i \left(\frac{\Lambda_{canopy,d} - R'_M}{G_T} \right) \Delta t \qquad (10.4)$$

where $W_{i,t}$ and $W_{i,t+\Delta t}$ are the weights (g dry matter m^2 ground) of a given plant part i (stem, roots, or storage organs) at time step t and subsequent time step $t+\Delta t$, respectively; Δt is the interval for each time step (day); $\Lambda_{canopy,d}$ is the daily gross canopy CO_2 assimilation (g CH_2O m^2 ground day^{-1}) (Eq. 7.23); R'_M is the maintenance respiration requirement (g CH_2O m^2 ground day^{-1}); G_T is the total glucose requirement (g CH_2O g^{-1} dry matter); and F_i is the fraction of dry matter of plant part i to the whole plant (Table 10.5).

Note: the weight gain of leaves is calculated differently due to leaf death.

Equations

10.1.5 Leaf death

$$\varepsilon_L = MAX\left(\varepsilon_{age}, \varepsilon_{sh}\right) \tag{10.5a}$$

where ε_L is the actual death rate of leaves (day^{-1}); and MAX function returns the largest of the enclosed values. As shown above, there are two possible reasons for leaf death: due to leaf age (ε_{age}) and due to self-shading of leaves (ε_{sh}).

Leaf death by age is determined by

$$\varepsilon_{age} = \begin{cases} 0 & (2-\xi_s) \geq 1.0 \\ \xi_r/(2-\xi_s) & 0.1 < (2-\xi_s) < 1.0 \\ \xi_r/0.1 & (2-\xi_s) \leq 0.1 \end{cases} \tag{10.5b}$$

where ε_{age} is the leaf death rate due to leaf age (day^{-1}); and ξ_r and ξ_s are the growth development rate (day^{-1}) and growth development stage (unitless), respectively.

Leaf death by self-shading is determined by

$$\varepsilon_{sh} = \begin{cases} 0 & L \leq L_{max} \\ 0.03 \times MIN\left[1, \dfrac{(L-L_{max})}{L_{max}}\right] & L > L_{max} \end{cases} \tag{10.5c}$$

where ε_{sh} is the leaf death rate due to leaf self-shading (day^{-1}); L and L_{max} are the leaf area index (LAI) and maximum LAI, respectively (m^2 leaf m^{-2} ground); and MIN is function to determine the smallest of the enclosed values.

10.1.6 Weight of leaves

10.1.6.1 Dead leaves

$$W_{DL,t+\Delta t} = W_{DL,t} + \left(W_{GL,t}\varepsilon_{L,t}\right)\Delta t \tag{10.6}$$

256

where $W_{DL,t}$ and $W_{DL,t+\Delta t}$ are the weights (g dry matter m^{-2} ground) of dead leaves at time step t and subsequent time step $t+\Delta t$, respectively; Δt is the interval for each time step (days); and $\varepsilon_{L,t}$ is the death rate of leaves at time step t (day^{-1}).

10.1.6.2 Green leaves

$$W_{GL,t+\Delta t} = W_{GL,t} +$$
$$\left[F_{GL} \left(\frac{\Lambda_{canopy,d} - R'_M}{G_T} \right) - W_{GL,t}\varepsilon_{L,t} \right] \Delta t \qquad (10.7)$$

where $W_{GL,t}$ and $W_{GL,t+\Delta t}$ are the weights (g dry matter m^{-2} ground) of green leaves at time step t and subsequent time step $t+\Delta t$, respectively; Δt is the interval for each time step (day); $\varepsilon_{L,t}$ is the death rate of leaves at time step t (day^{-1}); $\Lambda_{canopy,d}$ is the daily gross canopy CO_2 assimilation (g CH_2O m^{-2} ground day^{-1}) (Eq. 7.23); R'_M is the maintenance respiration requirement (g CH_2O m^{-2} ground day^{-1}); G_T is the total glucose requirement (g CH_2O g^{-1} dry matter); and F_{GL} is the dry matter fraction of green leaves to the whole plant (Table 10.5).

10.1.7   Leaf area index

$$L = W_{GL} \cdot SLA \qquad (10.8)$$

where L is the leaf area index (m^2 leaf m^{-2} ground); W_{GL} is the weight of green leaves (g dry matter m^{-2} ground); and SLA is the specific leaf area (m^2 leaf g^{-1} leaf dry matter) (Table 10.6).

Table 10.6. Specific leaf area (*SLA*) according to the crop growth stage (ξ_s).

ξ_s	SLA (m² leaf g⁻¹ leaf dry matter)
0.0	0.035
0.2	0.033
0.4	0.030
0.6	0.028
0.8	0.025
1.0	0.023
2.0	0.010

10.1.8 Mean leaf width

$$w_{t+\Delta t} = MIN\left[w_m, \left(w_t + w_g \Delta t \right) \right] \qquad (10.9)$$

where w_t and $w_{t+\Delta t}$ are the mean leaf width (m) at time step t and $t+\Delta t$, respectively; w_m is the maximum leaf width (m); w_g is the mean leaf width expansion rate (m day⁻¹); Δt is the time step interval (day); and MIN is the function to determine the smallest of the enclosed values.

10.1.9 Plant height

$$h_{t+\Delta t} = h_t + \frac{dh}{dt} \Delta t \qquad (10.10a)$$

$$\frac{dh}{dt} = T_{ts} \cdot \frac{b_0 b_1 h_m \exp\left(-b_1 T_{\Sigma ts}\right)}{\left[1 + b_0 \exp\left(-b_1 T_{\Sigma ts}\right) \right]^2} \qquad (10.10b)$$

where h_t and $h_{t+\Delta t}$ are the plant heights (m) at time t and subsequent time step $t+\Delta t$, respectively; Δt is the interval for each time step (day); h_m is the maximum possible height of the plant (m); and b_0 and b_1 are the intercept (unitless) and slope ($°C^{-1}$ day^{-1}) coefficients, respectively; T_{ts} is the daily temperature sum (°C day); and $T_{\Sigma ts}$ is the cumulative temperature sum (°C day).

The daily temperature sum is calculated by

$$T_{ts} = \begin{cases} 0 & T_{mean} \leq T_b \\ T_{mean} - T_b & T_{mean} > T_b \end{cases} \tag{10.11}$$

where T_{ts} is the daily temperature sum (°C day); T_b is the crop's base temperature (°C) below which crop growth ceases (Table 10.2); and T_{mean} is the mean of the minimum and maximum air temperature (°C).

The summation of each day's temperature sum gives the cumulative temperature sum, and it is calculated by

$$T_{\Sigma ts} = \sum_t T_{ts,t} \tag{10.12}$$

where $T_{\Sigma ts}$ is the cumulative temperature sum (°C day); and $T_{ts,t}$ is the temperature sum at day t (°C day).

10.1.10 Rooting depth

$$d_{\text{root},t+\Delta t} = MIN\left[d_m, \left(d_{\text{root},t} + d_g \Delta t\right)\right] \tag{10.13a}$$

$$d_g = \begin{cases} 0 & \text{if } \left(\Theta_{v,t} \leq \Theta_{v,wp}\right) \text{ or } \left(\xi_s \geq 1\right) \\ d_g & \text{if } \left(\Theta_{v,t} > \Theta_{v,wp}\right) \text{ and } \left(\xi_s < 1\right) \end{cases} \tag{10.13b}$$

where $d_{\text{root},t}$ and $d_{\text{root},t+\Delta t}$ are the rooting depth (m) at time step t and $t+\Delta t$, respectively; d_m is the maximum rooting depth (m); d_g is the root elongation rate, denoting rooting depth increase per day (m day^{-1}); Δt is the time step interval (day); ξ_s is the growth development stage; $\Theta_{v,t}$ is the volumetric soil water content at time t (m^3 m^{-3}); $\Theta_{v,wp}$ is volumetric soil water content at permanent wilting point (m^3 m^{-3}); and MIN is the function to determine the smallest of the enclosed values.

Eq. 10.13b shows that root elongation only occurs when the soil water content is above the permanent wilting point and the growth development stage is before anthesis.

Note: no root death is assumed.

10.1.11 Reduction due to water stress

Insufficient water limits growth according to the ratio between actual and potential plant transpiration (Eq. 9.4):

$$R_{D,T} = \frac{AET_c}{PET_c}$$

where $R_{D,T}$ is the reduction factor for growth (ranging from 0 for zero growth to 1 for no growth suppression due to water stress); and AET_c and PET_c are the actual (Eq. 9.4) and potential (Eq. 8.24) transpiration rates (mm day^{-1}), respectively.

The amount of assimilates potentially produced (*i.e.*, in the absence of water stress) is reduced in proportion to the level of water stress by

$$\Lambda'_{canopy,d} = \Lambda_{canopy,d} \cdot R_{D,T} \qquad (10.14)$$

and plant height growth rate is reduced to

$$\frac{dh'}{dt} = \frac{dh}{dt} \cdot R_{D,T} \qquad (10.15)$$

as well as mean leaf width and root elongation rate to

$$w'_g = w_g \cdot R_{D,T} \tag{10.16}$$

$$d'_g = d_g \cdot R_{D,T} \tag{10.17}$$

10.2 Measured parameters

The following are parameters that must be measured or supplied to the crop growth development model component:

1. The dependence of crop growth development rate (ξ_r) on air temperature (Table 10.2).

2. The crop's base temperature (T_b) both before and after anthesis as 0 °C.

3. The maintenance respiration coefficient (k_i) and glucose requirement (G_i) for the growth of the various plant parts (Table 10.3 and 10.4).

4. The dry matter partitioning (F_i) of the various plant parts at different stages of crop growth (Table 10.5).

5. The specific leaf area (SLA) at different stages of crop growth (Table 10.6).

6. The maximum rooting depth (d_m) and the root elongation rate (d_g) are set at 0.6 m and 0.012 m day^{-1}, respectively. It is important that the maximum rooting depth do not exceed the total soil profile depth (*i.e.*, cumulative thickness of the last soil layer, S_N) we had set in the previous chapter (*see* section 9.2.3).

7. The maximum leaf width (w_m) is set to 0.08 m and the expansion rate of the mean leaf width (w_g) as 0.0023 m day^{-1}.

8. The maximum plant height (h_m) is set at 0.4 m and the plant height coefficients b_0 and b_1 at 10 and 0.008, respectively.
9. Initial values:
   a) Initial growth development stage (ξ_s) is set at 0.1.
   b) Initial plant part dry weight (g m^{-2} ground) for green leaves (W_{GL} = 1.44), dead leaves (W_{DL} = 0), stem (W_S = 0.29), roots (W_R = 0.16), and storage organs (W_O = 0).
   c) Both initial plant height and rooting depth set at 0.04 m.
   d) Initial mean leaf width is set at 0.008 m.

10.3 Implementation

Create a new worksheet in the gcg model workbook, and name this new worksheet *Growth* (Fig. 10.2).

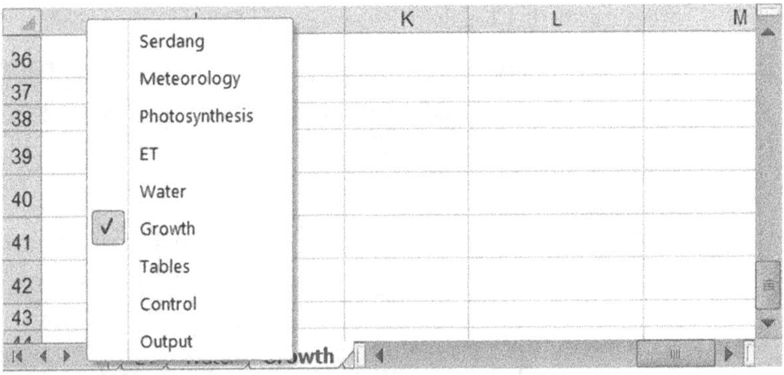

Fig. 10.2. Insert a new worksheet in the gcg workbook. This new worksheet, named *Growth*, will contain all equations in the crop growth development model component.

This brings the total number of worksheets in the gcg workbook to nine. The Growth worksheet will contain the implementation of the crop growth development model component based on Eq. 10.1 to 10.17.

We will also continue to define new cell names for improving model readability and model documentation purpose. These new cell names are as listed in Table 10.7.

Table 10.7. Cell names defined for the crop growth development model component, in addition to those already defined for the previous model components (Table 6.2, 7.3, 8.1, and 9.2).

Worksheet	Cell	Cell name
Control	G32	option
Growth	B19	dm
	B20	dg
	B23	hm
	B24	h_b0
	B25	h_b1
	B28	wm
	B29	wg
	B33	kmGL
	B34	kmS
	B35	kmR
	B36	kmO
	B39	GGL
	B40	GS
	B41	GR
	B42	GO
	E2	dvs
	E5	dvr
	E7	dayassim_c
	E15	Tb

263

Worksheet	Cell	Cell name
	E16	`Tts`
	E17	`Tts_total`
	H2	`WGL`
	H3	`WDL`
	H4	`WS`
	H5	`WR`
	H6	`WO`
	H7	`WT`
	H11	`RM_c`
	H20	`eage`
	H23	`esh`
	H24	`eL`
	H27	`SLA`
	I7	`RM`
	J2	`FGL`
	J4	`FS`
	J5	`FR`
	J6	`FO`
	K7	`GT`
Tables	G4:G6	`dvr_Ta`
	G13:G20	`frac_dvs`
	H4:H6	`dvr_dvr0`
	H13:H20	`frac_FGL`
	I4:I6	`dvr_dvr1`
	I13:I20	`frac_FS`
	J13:J20	`frac_FR`
	K13:K20	`frac_FO`
	L13:L20	`frac_SLA`

The tabulated values from Table 10.2, 10.5, and 10.6 are added to the Tables worksheet (Fig. 10.3). Selected cell ranges are given cell names such as cells G4:G6 with dvr_Ta, H4:H6 with dvr_d0 (pre-anthesis ξ_r values), and I4:I6 with dvr_d1 (post-anthesis ξ_r values) to simplify reference to them (Table 10.7).

	F	G	H	I	J	K	L	M
1								
2			ξ_r					
3		T_a (°C)	$\xi_s < 1$	$\xi_s \geq 1$				
4		0	0	0				
5		30	0.027	0.031				
6		40	0.027	0.031				
7								
8								
9								
10								
11		FRACTION OF PLANT PARTS AND SLA						
12		ξ_s	F_{GL}	F_S	F_R	F_O	SLA	
13		0	0.76	0.15	0.09	0	0.035	
14		0.2	0.67	0.21	0.12	0	0.033	
15		0.4	0.66	0.29	0.05	0	0.03	
16		0.6	0.62	0.34	0.04	0	0.028	
17		0.8	0.38	0.53	0.09	0	0.025	
18		1	0.27	0.51	0.17	0.05	0.023	
19		2	0	0	0	1	0.01	
20		2.2	0	0	0	1	0.01	
21								

Fig. 10.3. Tables worksheet: tabulated values to determine the growth development rate (ξ_r), fraction of various plant parts (F_i), and specific leaf area (SLA). These tabulated values are from Table 10.2, 10.5, and 10.6.

Fig. 10.4 shows the list of required model inputs which includes the maintenance (cells B33:B36) and glucose growth requirement (cells B39:B42) for the various plant parts, taken from Table 10.3 and 10.4, respectively.

a)

	A	B	C
1	INPUT		
2	*Initial weights*		
3	W_{GL}	1.44	
4	W_{DL}	0	
5	W_S	0.29	
6	W_R	0.16	
7	W_O	0	
8			
9	Initial growth stage	0.1	
10	Initial mean leaf width	0.008	
11	Initial height	0.04	
12	Initial rooting depth	0.04	
13			
14	*Base temperature*		
15	$T_b\ (\xi_s<1)$	0	
16	$T_b\ (\xi_s\geq1)$	0	
17			
18	*Roots*		
19	d_m	0.6	
20	d_g	0.012	
21			

b)

	A	B	C
22	*Plant height*		
23	h_m	0.4	
24	intercept, b_0	10	
25	slope, b_1	0.008	
26			
27	*Leaf width*		
28	w_m	0.08	
29	w_g	0.0023	
30			
31	CONSTANTS		
32	*Maintenance*		
33	k_{GL}	0.03	
34	k_S	0.015	
35	k_R	0.015	
36	k_O	0.01	
37			
38	*Growth*		
39	G_{GL}	1.463	
40	G_S	1.513	
41	G_R	1.444	
42	G_O	1.415	
43			

Fig. 10.4. Growth worksheet: model inputs (a and b) for the growth development model component.

Fig. 10.5 shows the calculations for the growth development stage (cells E2:E5), mean leaf width (cells E11:E12), plant height (cells E15:E22), and rooting depth (cells E25:E29). These calculations are based on Eq. 10.1, 10.9, 10.10, and 10.13, respectively.

	C	D	E	F
1		**DEV. STAGE**		
2		ξ_s		
3		ξ_r (ξ_s<1)	=interpolate(dvr_Ta,dvr_dvr0,Tmean)	
4		ξ_r ($\xi_s \geq 1$)	=interpolate(dvr_Ta,dvr_dvr1,Tmean)	
5		ξ_r	=IF(dvs<=1,E3,E4)	
6				
7		Λ'_{canopy}	=dayassim*RDT	
8				
9		**LEAF WIDTH**		
10		w (UPD)		
11		w	=MIN(wm,E10)	
12		w'_g	=wg*RDT	
13				
14		**PLANT HEIGHT**		
15		T_b	=IF(dvs<=1,B15,B16)	
16		T_{ts}	=MAX(0,Tmean-Tb)	
17		$T_{\Sigma ts}$		
18		h (UPD)		
19		h	=MIN(hm,E18)	
20		n7	=h_b0*h_b1*hm*EXP(-h_b1*Tts_total)	
21		n8	=(1+h_b0*EXP(-h_b1*Tts_total))^2	
22		dh'/dt	=Tts*E20/E21*RDT	
23				
24		**ROOT DEPTH**		
25		d_{root} (UPD)		
26		d_{root}	=MIN(dm,E25)	
27		(ξ_s<1)?	=dvs<1	
28		($\Theta_v > \Theta_{wp}$) ?	=Water!E12>Water!E9	
29		d'_g	=IF(AND(E27,E28),dg,0)*RDT	
30				

Fig. 10.5. Growth worksheet: calculations for the growth development stage, mean leaf width, plant height, and rooting depth.

Cell E5 uses the IF function to determine the correct growth development rate (ξ_r) to use depending on the current growth stage (whether before or after anthesis). The current growth development rate is read from the table we had prepared earlier in the Tables worksheet (cells Tables!G4:I6). BuildIt's interpolate function is used for linear interpolation to determine the growth development stage based on daily mean air temperature.

Fig. 10.5 also shows that the water stress level ($R_{D,T}$ or cell Water!E13) will reduce the mean leaf width expansion (cell E12), plant height growth (cell E22), and rooting depth (cell E29) accordingly as indicated in Eq. 10.15 to 10.17. The daily canopy photosynthesis is also reduced by water stress according to Eq. 10.14 in cell E7.

Fig. 10.6 and 10.7 show the calculations to determine the dry weights of the various plant parts. Cells I2:I7 and K2:K7 implement Eq. 10.2b and 10.3, respectively. BuildIt's interpolate function is used in cells J2:J7 to determine, by linear interpolation if necessary, the fraction of a given plant part to the whole plant based on the current growth development stage. The last column, cells L2:L6, determine the growth rate of the plant parts based on Eq. 10.4, 10.6, and 10.7.

Cell H11 (Fig. 10.7) uses the MIN function to ensure that the maintenance requirement never exceeds the daily photosynthetic rate. If it does, the assimilates available for growth is zero (*i.e.*, all the assimilates used for maintenance). Consequently, the IF function is used in cell H12 to ensure no division by zero occur should assimilates for growth is ever zero. Cells H16:H24 and H27:H28 implement Eq. 10.5 and 10.8, respectively.

	F	G	H	I	J	K	L	M
1		Part i	W_i	k_iW_i	F_i	F_iG_i	Growth rate	
2		W_{GL}		=WGL*kmGL	=interpolate(frac_dvs,frac_FGL,dvs)	=FGL*GGL	=FGL*H13-WGL*eL	
3		W_{DL}					=WGL*eL	
4		W_S		=WS*kmS	=interpolate(frac_dvs,frac_FS,dvs)	=FS*GS	=FS*H13	
5		W_R		=WR*kmR	=interpolate(frac_dvs,frac_FR,dvs)	=FR*GR	=FR*H13	
6		W_O		=WO*kmO	=interpolate(frac_dvs,frac_FO,dvs)	=FO*GO	=FO*H13	
7		Total		=SUM(I2:I6)		=SUM(K2:K6)		
8								

Fig. 10.6. Growth worksheet: calculations to determine the plant part dry weights.

	F	G	H
9		**RESPIRATION**	
10		n1	=RM*2^((Tmean-25)/10)*WGL/(WGL+WDL)
11		R'_M	=MIN(dayassim_c,H10)
12		$\Lambda'_{canopy} - R'_M$	=dayassim_c-RM_c
13		$(\Lambda'_{canopy}-R'_M) / G_T$	=IF(GT>0,H12/GT,0)
14			
15		**LEAF DEATH**	
16		$(2-\xi_s)$	=2-dvs
17		n2	0
18		n3	=dvr/H16
19		n4	=dvr/0.1
20		ε_{age}	=IF(H16>=1,H17,IF(H16<=0.1,H19,H18))
21		n5	0
22		n6	=0.03*MIN(1,(L-Lmax)/Lmax)
23		ε_{sh}	=IF(L<=Lmax,H21,H22)
24		ε_L	=MAX(eage,esh)
25			
26		**LEAF AREA**	
27		SLA	=interpolate(frac_dvs,frac_SLA,dvs)
28		L	=WGL*SLA

Fig. 10.7. Growth worksheet: calculations for the growth and maintenance respiration rates, as well as the leaf death and leaf area expansion rates.

Notice from Fig. 10.5 and 10.6 that cells E2 (growth development stage), E10 (mean leaf width), E18 (plant height), E25 (rooting depth), and cells H2:H6 (plant part dry weights) are blank. This is because we will use BuildIt's INI action (defined in the prerun section) to initialize their values (Fig. 10.8), after which we will use the UPD action (defined in the operation section) to update their values based on their current growth rates (Fig. 10.9).

	F	G	H	I	J	K
19						
20		**PRERUN**				
21		INI	=Water!H8:J8	=Water!H7:J7		
22		INI	=Growth!H2:H6	=Growth!B3:B7		
23		INI	=dvs	=Growth!B9		
24		INI	=Growth!E10	=Growth!B10		
25		INI	=Growth!E18	=Growth!B11		
26		INI	=Growth!E25	=Growth!B12		
27						
28						
29						
30						
31		**OPTION**				
32		RUN	ClearOutput			
33						

Fig. 10.8. Control worksheet: The prerun section (cell G21 onwards) contains several INI actions to initialize the model parameters and the option section (cell G32 onwards) contains the RUN action to execute the ClearOutput macro to delete the previous model output listing, if any.

	F	G	H	I	J	K	L	M	N	O
1		**OPERATION**								
2		ITG	=th	=assim	=Photosynthesis!J20	=tsr	=tss			
3		ITG	=th	=LETs	=ET!H20	0	12			
4		ITG	=th	=LETc	=ET!H26	0	12			
5		UPD	=Water!H8:J8	=Water!H22:J22	100					
6		UPD	=Growth!H2:H6	=Growth!L2:L6						
7		UPD	=Growth!E10	=Growth!E12						
8		ACC	=Tts	=Tts_total	+					
9		UPD	=Growth!E18	=Growth!E22						
10		UPD	=Growth!E25	=Growth!E29						
11		UPD	=dvs	=dvr						
12										

Fig. 10.9. Control worksheet: several UPD actions are defined to update the model parameters. The ACC action in row 8 is to accumulate the temperature sum values by summation to obtain the cumulative temperature sum.

It is crucial that the growth development stage be updated only after all calculations for the current time step have been completed. This is why the UPD action to update the growth development stage is placed last in the operation section (row 11 in Fig. 10.9).

All the previous calculations depend on the current growth development stage, so if we were to update the growth stage too early, it risks some calculations erroneously using the future (and not the current) growth stage value.

Also notice that cell E17 is blank (Fig. 10.5). Temperature sum is calculated in cells E15 and E16 (Eq. 10.11), after which the result in E16 is added to the current value in cell E17; thus, updating the value in E17 to give the cumulative temperature sum (Eq. 10.12).

To calculate the cumulative temperature sum, we use the ACC action (*see* section 3.5.1). This action is defined in row 8 in the operation section in the Control worksheet (Fig. 10.9). Note the + operator in cell J8 in the Control worksheet to instruct that the accumulation of values is by summation.

Prior to the implementation of the crop growth development model component, provisional values were given for leaf area index (Fig. 7.4), plant height and mean leaf width (Fig. 8.5), and rooting depth (Fig. 9.5).

All these parameters can now be modeled with the implementation of this crop growth development model component.

So, we should remove these provisional values, as shown in Fig. 10.10 to 10.12.

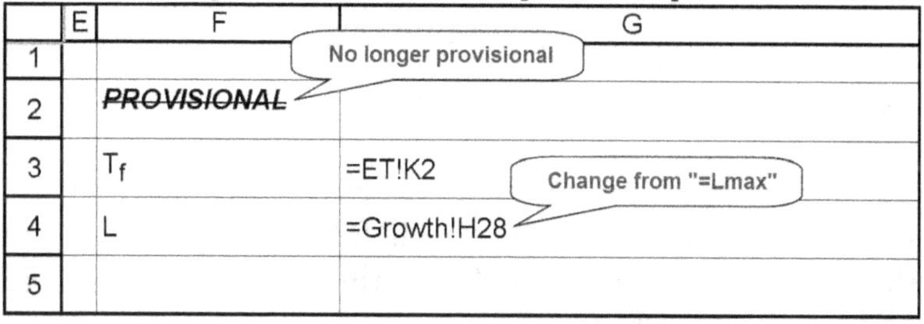

	E	F	G
1			No longer provisional
2		*PROVISIONAL*	
3		T_f	=ET!K2 — Change from "=Lmax"
4		L	=Growth!H28
5			

Fig. 10.10. Photosynthesis worksheet: cell G4 is modified so it refers to cell H28 in the Growth worksheet for the leaf area index value. Cell G4 is still defined with the cell name L (Table 7.3).

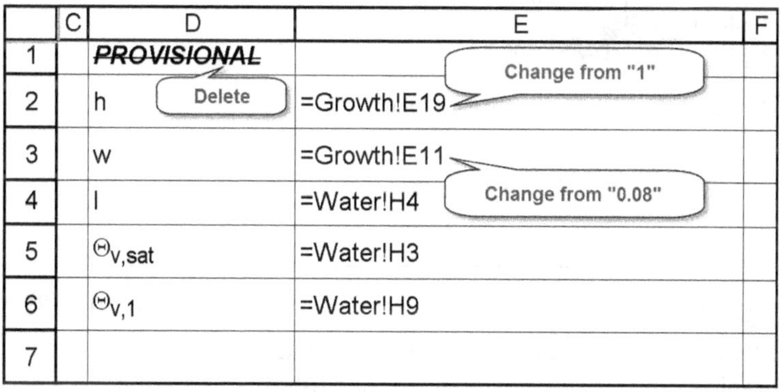

	C	D	E	F
1		*PROVISIONAL*		
2		h — Delete	=Growth!E19 — Change from "1"	
3		w	=Growth!E11	
4		l	=Water!H4 — Change from "0.08"	
5		$\Theta_{v,sat}$	=Water!H3	
6		$\Theta_{v,1}$	=Water!H9	
7				

Fig. 10.11. ET worksheet: cells E2 (for plant height) and E3 (for the mean leaf width) are modified so they refer to cells E19 and E11 in the Growth worksheet, respectively. Cells E2 and E3 are still defined with the cell names h and leafwidth, respectively (Table 8.1).

	C	D	Delete	E		F
1		*PROVISIONAL*		Change from "0.3"		
2		d_{root}	=Growth!E26			
3						

Fig. 10.12. Water worksheet: cell E2 refers to cell E26 in the Growth worksheet for the rooting depth value. Cell E2 is still defined with the cell name `droot` (Table 9.2).

The loop information in the Control worksheet is modified slightly so that simulation of the crop growth and yield ends when the crop reaches growth stage 2 or the maturity stage (Fig. 10.13).

	A	B	C	D	E	F
1	INPUT					
2	date	=DATE(2000,1,1)		current date	=B2+INT(_step)	
3				doy	=date-DATE(YEAR(date),1,0)	
4	CONTROL			hour		
5						
6	stepsize	1				
7	step					
8	criteria	=dvs<=2				
9						

Fig. 10.13. Control worksheet: loop criteria (cell B8) is modified so that simulation runs until the growth development stage reaches stage 2 (plant maturity).

The Output worksheet is also modified to change the output listing to include the number of days elapsed, growth development stage, plant part dry weights, plant height, rooting depth, leaf area index, and the total water content in the whole soil profile (Fig. 10.14). Cell A3 calculates the number of days elapsed since the start of the simulation date (Jan. 1, 2000 in this case).

	A	B	C	D	E	F	G	H	I	J	K	L
1	TO OUTPUT											
2	days	ξ_s	W_{GL}	W_{DL}	W_s	W_R	W_O	h	d_{root}	L	$\Theta_{v,root}$	
3	=date-Control!B2	=dvs	=WGL	=WDL	=WS	=WR	=WO	=h	=droot	=L	=Water!E12	
4		FALSE	FALSE	FALSE	FALSE	FALSE	FALSE	FALSE	FALSE	FALSE	FALSE	
5												
6												
7	OUTPUT											
8												
9												

Fig. 10.14. Output worksheet: the model output will include the number of days elapsed, growth development stage, plant part dry weights, plant height, rooting depth, leaf area index, and the total water content in the whole soil profile. Cell A3 is defined with cell name `_read` and cell A8 `_write`. Note: ensure cells in column A have been formatted to display in General (and not in the default Date) format (*see* Fig. 6.15).

The FALSE values in cells B4:K4 (Fig. 10.14) are to ensure the corresponding parameter values are outputted before their values are changed in the operation section.

Lastly, we created the option section (Fig. 10.8) by giving cell G32 in the Control worksheet the cell name _option (Table 10.7). Here, we defined a single RUN action to execute the ClearOutput macro (*see* section 5.2) to clear any model output from the previous model run. Creating this option section to run the ClearOutput macro is optional but is convenient to have because deleting the previous model output can eliminate any overlap of model results from two or more previous model runs.

To start the simulation, click the "Start Simulation" from the BuildIt menu. The model output is as shown in Fig. 10.15 and the charts drawn from the model output in Fig. 10.16.

	A	B	C	D	E	F	G	H	I	J	K	L
1	TO OUTPUT											
2	days	ξ_s	W_{GL}	W_{DL}	W_S	W_R	W_O	h	d_{root}	L	$\Theta_{v,root}$	
3	0	0	0	0	0	0	0	0.4	0.6	0	0.01	
4		FALSE	FALSE	FALSE	FALSE	FALSE	FALSE	FALSE	FALSE	FALSE	FALSE	
5												
6												
7	OUTPUT											
8	0	0.1	1.44	0	0.29	0.16	0	0.04	0.04	0.04896	0.2	
9	1	0.125	1.864	0	0.397	0.222	0	0.049	0.052	0.06293	0.254	
10	2	0.149	2.31	0	0.515	0.291	0	0.059	0.064	0.07742	0.272	
11	3	0.173	2.94	0	0.692	0.393	0	0.07	0.076	0.0978	0.284	
12	4	0.197	3.588	0	0.884	0.503	0	0.084	0.088	0.11849	0.267	
13	5	0.221	4.406	0	1.139	0.649	0	0.098	0.1	0.14401	0.29	
73	65	1.793	54.45	172.9	264	68.69	136.1	0.4	0.496	0.6908	0.269	
74	66	1.823	46.98	180.8	264.8	68.94	141.8	0.4	0.496	0.57786	0.262	
75	67	1.854	39.18	188.9	265.4	69.14	147.2	0.4	0.496	0.46637	0.256	
76	68	1.883	31.48	196.8	265.8	69.28	152.1	0.4	0.496	0.36256	0.25	
77	69	1.913	23.55	204.9	266.1	69.37	155.9	0.4	0.496	0.26207	0.245	
78	70	1.943	16.61	211.9	266.2	69.41	158.7	0.4	0.496	0.17842	0.3	
79	71	1.973	11.71	216.8	266.3	69.43	161.1	0.4	0.496	0.12122	0.282	
80												

Fig. 10.15. Output worksheet: the model output. Rows 14 to 72 are hidden to show the start and end of the output list in a single screenshot.

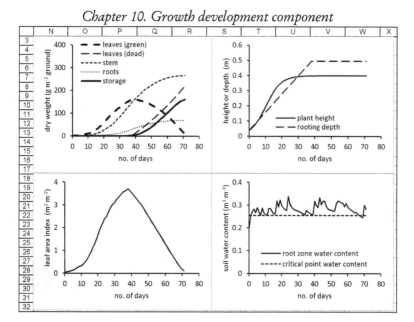

Fig. 10.16. Output worksheet: charts are drawn from the model output to visually depict the change in the plant part dry weights, plant height, rooting depth, leaf area index, and the total soil water content.

The model output shows that crop reached growth stage 1 and 2 in 35 and 71 days, respectively. The crop reached a maximum leaf area index of 3.7 m^2 leaf m^{-2} ground at growth stage 1, after which it declined in a near linear manner. The maximum dry weight for green leaves was 163 g leaf m^{-2} ground. The maximum plant height achieved was 0.4 m and rooting depth was 0.5 m in approximately 30 and 40 days, respectively. Due to heavy and frequent rains in Serdang, the crop very rarely suffered any water stress. The total soil water content was almost always above the critical soil water level, a level below which the crop's growth would be reduced due to the effects of water stress.

277

Lastly, Fig. 10.17 shows the exchange of information between the five gcg model components.

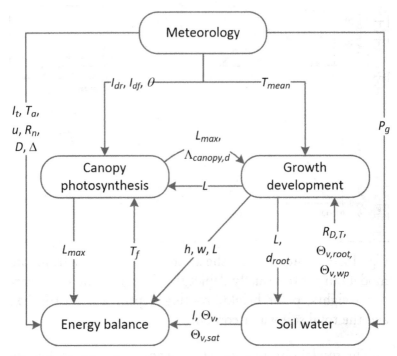

Fig. 10.17. The flow (exchange) of information between the gcg model components. Key: I_t, I_{dr}, and I_{df} = total, direct, and diffuse solar irradiance, respectively; T_a, T_f, and T_{mean} = current temperature of the air and canopy and the mean air temperature, respectively; u = wind speed; R_n = net radiation; D = vapor pressure deficit; Δ = slope of the saturated vapor pressure curve; P_g = gross rainfall; θ = solar inclination; $\Lambda_{canopy,d}$ = daily canopy photosynthesis; L_{max} and L = maximum and current leaf area index, respectively; d_{root} = rooting depth; h = plant height; w = mean leaf width; l = thickness of the uppermost soil layer; $R_{D,T}$ = water stress level; $\Theta_{v,root}$ = volumetric soil water content in the rooting zone; $\Theta_{v,wp}$ and $\Theta_{v,sat}$ = volumetric soil water content at permanent wilting and saturation point, respectively; and Θ_v = current volumetric soil water content.

10.4 Exercises

1. Modify the model so that the model can be used to determine the effects of higher CO_2 concentration levels and air temperature on the growth and yield of the crop. The following simulation conditions to run the model are as follows:

   a) Intercellular CO_2 concentration, C_i, increasing to 350 then 430 µmol mol^{-1}.

   b) Air temperature, T_a, increasing by 1 then 2 °C.

   c) Both (a) and (b) conditions.

   Run the model for Serdang starting from Jan. 1, 2000. Examine the model output from the various scenarios to determine the individual and combined effects of higher CO_2 concentration levels and air temperature on the growth and yield of the crop, particularly on the daily photosynthetic rate, growth and yield parameters (weight of leaves, stem, and roots, plant height, and rooting depth, and the total leaf area), and the length of growing period (*i.e.*, time it takes to reach $\xi_s = 2$). Make all comparisons against the baseline which is the model output when C_i is 270 µmol mol^{-1} and T_a is as calculated from Eq. 6.17.

2. Obtain the weather data for one year for your site, and run the model for two different dates, one in the beginning and the other in the middle of the year, to determine the differences in the crop growth as well as the length of growing period between these two dates.

Explain why there are differences (or similarities) in the crop growth and length of growing period between these two dates.

Hint: compare between dates their average daily air temperature and solar irradiance.

3. Suppose that when the crop reaches the flowering stage (ξ_s = 1), attack from pests would reduce the leaf dry weight by 50%. Modify the model to account for this scenario. Run the model to determine the effect on crop growth due to this pest attack.

   Note: this reduction in leaf dry weight by 50% would only occur *once* when pests attacked the crop at ξ_s = 1. After ξ_s = 1, attack from pests would cease.

4. From Question 3: suppose instead that the attack from pests would occur earlier; that is, when the crop reaches growth stage ξ_s = 0.5. The pests would reduce the leaf dry weight also by 50%. Run the model again but for this earlier pest attack.

   Which period of pest attack (at ξ_s = 0.5 vs. ξ_s = 1) would have a larger detrimental impact on plant growth?

5. Modify the model to reduce the daily rainfall amount by 50%. This would increase the occurrence and severity of water stress crop growth. Run the model to determine the impact of 50% less rainfall amount on crop growth.

PART III
Relationships between variables and formulas

Chapter 11. Cells network map

11.1 Visualizing the cells interrelationships

Excel provides two utilities, *Trace Precedents* and *Trace Dependents*, that allow us to determine a formula's precedent and dependent cells.

Let's say cell D1 contains the formula: "=P1" which means cell D1 requires information from cell P1, so cell D1 is dependent on cell P1. In contrast, cell P1 provides information to cell D1, so cell P1 is the precedent of cell D1.

BuildIt supplements these two Excel's tracing utilities by providing a tool called *Trace*. Trace also depicts the relationships between spreadsheet cells but in a much more visual manner by drawing a cells network map that shows the flow of information from one cell to another (Fig. 11.1).

Trace can be used as a visual tool to help in better understanding a model, to detect model implementation errors, and to discover direct and indirect relationships between variables and formulas.

Fig. 11.1 is such a cells network map produced by Trace. This map depicts the interrelationships between all cells in a worksheet called *Calculations* (Fig. 11.2a). Note that Trace uses the worksheet name as the title of the map.

Trace represents cells by a variety of shapes depending on what these cells contain. Trace uses: a) circles for cells containing values (*e.g.*, cell B1), b) rectangles for formulas (*e.g.*, cells B2, B3, and B4), c) parellograms for a range of cells (*e.g.*, cell range B1:B3), and d) triple-layered octagons for external cells, which are cells that belong in another worksheet whether in the current or in another workbook.

Visualizing the cells interrelationships

In Fig. 11.1, cells Constant!B1 and Constant!B2 are from another worksheet named *Constants*; thus, from the perspective of the Calculations worksheet, both these cells are external cells, so they are represented as octagons.

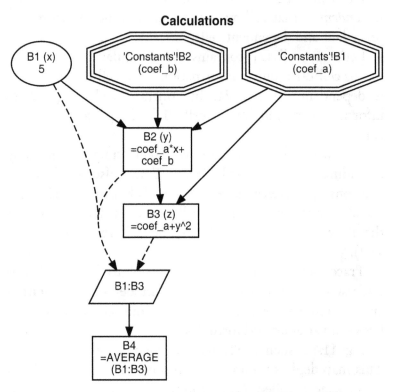

Fig. 11.1. A cells network map produced by Trace. This map shows how the various spreadsheet cells in the *Calculations* worksheet are related to one another.

Trace uses arrows to indicate the direction of information flow. In Fig. 11.1, we can see that the value 5 in cell B1 is used by cells B2 and B4. Cell B2 also requires the values from two external cells: Constants!B1 and Constants!B2.

284

Dashed arrows are drawn when a range of cells is referred by one or more cells. In Fig. 11.1 and 11.2, the cell range B1:B3 is used by cell B4. Consequently, a parallogram is drawn to represent this cell range B1:B3, and the dashed arrows link this cell range with its three cell members.

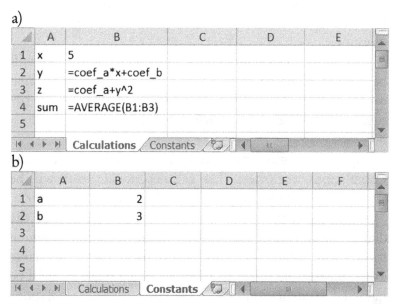

Fig. 11.2. A workbook with two worksheets, named a) *Calculations* and b) *Constants*. Trace is used to draw the cells network map for these two worksheets. In the Calculations worksheet, cells B1, B2, and B3 are given cell names x, y, and z, respectively, whereas in the Constants worksheet, cells B1 and B2 are given cell names coef_a and coef_b, respectively.

Trace has two options: *"Trace Workbook"* and *"Trace Selected Cells"* (Fig. A.3 and A.5). The "Trace Workbook" option draws the cells network map for all cells in one or more selected worksheets. Fig. 11.1, for instance, was produced by choosing the "Trace Workbook" command from BuildIt's menu.

On the other hand, "Trace Selected Cells" option draws the cells network map for only the currently selected cells. This option is useful when we want to quickly determine the network between a few selected cells. "Trace Selected Cells" option also produces a smaller map (since it involves fewer cells and not all the cells in the worksheet) for us to examine.

When the "Trace Workbook" option is chosen, we will be shown a dialog box (Fig. 11.3) to obtain additional inputs such as selecting from a list the worksheets in the current workbook for which we want to draw the cells network map.

Fig. 11.1 was produced by selecting only the Calculations worksheet. The cell dependency map will be saved as a picture file in the default GIF format.

Click the large > button on the bottom right of the dialog box to choose the location in the computer to which to save this picture file. This picture file has the same name as the worksheet name. In this case, the picture file will be named "Calculations.gif" because the Calculations worksheet was selected for tracing. If we additionally select the Constants worksheet for tracing, two picture files will be produced: "Constants.gif" and "Calculations.gif". In other words, each selected worksheet has its own picture file.

Once done, click the OK button. Once tracing is completed, we will be presented with a file list. Double click the file "Calculations.gif", and the computer's default picture preview or editor program will open the cells network map (which should look like Fig. 11.1).

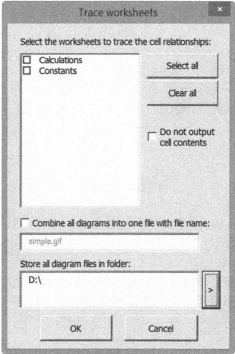

Fig. 11.3. Dialog box to prompt for user inputs when the "Trace Workbook" is chosen.

Notice that not all cells in the Calculations worksheet are included in Fig. 11.1. We can see that in Fig. 11.2a, cells A1 to A4 do not appear in Fig. 11.1 although these cells are not empty (*i.e.*, they contain text).

This is because Trace excludes cells that are neither precedent nor dependent cells. Only cells that provide information to other cells or require information from other cells are included by Trace. Cells A1 to A4 merely serve as text labels and do not have any relationships with other cells; thus, Trace ignores these four cells. Including them in the cells network map would only obfuscate the map especially if a large map is produced from a large and complex model.

That Trace ignores cells that are neither precedents nor dependents is why if we choose to draw the cells network map for the Constants worksheet, Trace will produce a seemingly odd output as shown in Fig. 11.4. This cells network map shows only one double-layered circle with the word *None*.

Constants

Fig. 11.4. An "uninteresting" cells network map for the Constants worksheet. Only one double-layered circle, with the word *None*, is drawn because, within the same worksheet, no cells interrelationships exist.

Fig. 11.2b shows that although the Constants worksheet is not empty, it does not have any cells that provide information to other cells or require information from other cells in the same worksheet. The only non-empty cells A1 and A2 contain only text and cells B1 and B2 values.

Without examining the Calculations worksheet, Trace would not know that cells B1 and B2 in the Constants worksheet are precedents to some cells in the Calculations worksheet (*see* Fig. 11.1 and Fig. 11.2a). Consequently, Trace considers the Constants worksheet, in isolation, as having no cell relationships.

We can however combine the cells network maps for two or more worksheets to produce a single, albeit larger, map.

Select both the Calculations and Constants worksheets from the worksheet list (Fig. 11.3). Place a checkmark next to the "Combine all diagrams into one file with file name:", and choose the desired file name for the combined cells network map. After clicking the OK button, Trace produces the combined cells map as shown in Fig. 11.5.

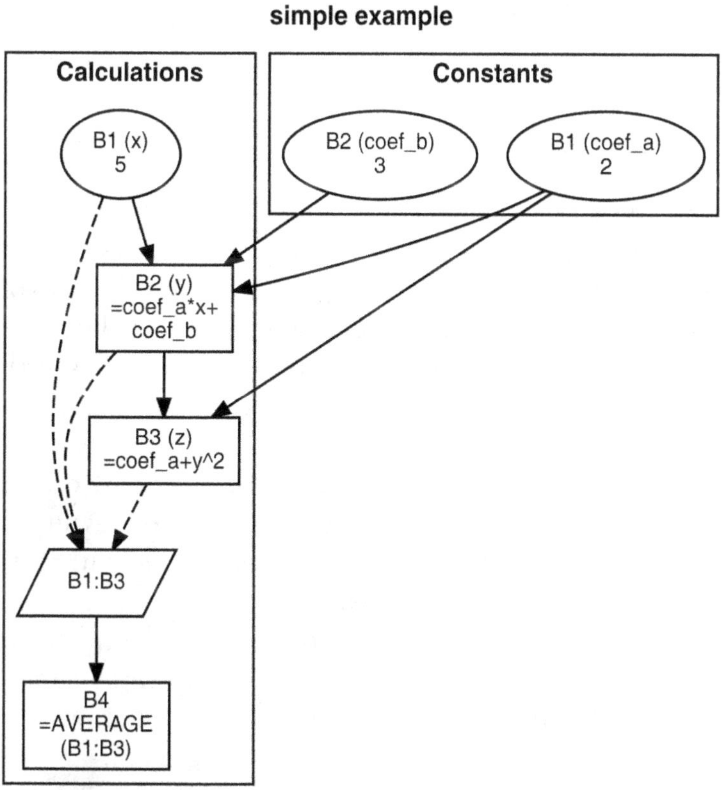

simple example

Fig. 11.5. A combined cells network map for the Calculations and Constants worksheets. The name of the workbook, *simple example*, is taken as the title of the combined map.

Two boxes are drawn, where each box represents a worksheet. The box entitled *Calculations* represents the cells network map for the Calculations worksheet, and the second box *Constants* the cells network map for the Constants worksheet.

As stated earlier, when Trace examines the Constants worksheet in isolation, Trace considers this worksheet as having no cells interrelationships, so Trace produces a cells network map like in Fig. 11.4. But when we now include the Calculations worksheet, Trace now sees that cells relations do exist between the two Constants and Calculations worksheets.

But why are cells B1 and B2 in the Constants worksheet drawn as single-layered circles in Fig. 11.5 but as octagons in Fig. 11.1? This is because Fig. 11.1 is the cells network map for only the Calculations worksheet, so both cells Constants!B1 and Constants!B2, seen from the perspective of the Calculations worksheet, are external cells. Fig. 11.5, on the other hand, simultaneously considers both the Calculations and Constants worksheets, so local cells B1 and B2 from the Constants worksheet are represented as single-layered circles by Trace and placed (drawn) in the map within the Constants box.

Trace uses the name of the workbook as the title of the combined cells network map. In this example, the name of the workbook which contains the Calculations and Constants worksheet is *simple example* (Fig. 11.5).

Lastly, Fig. 11.3 has an additional option: "Do not output cell contents". This option draws the cells network map but without showing the contents in the cells, such as shown in Fig. 11.6. This option is useful to reduce the size of the map.

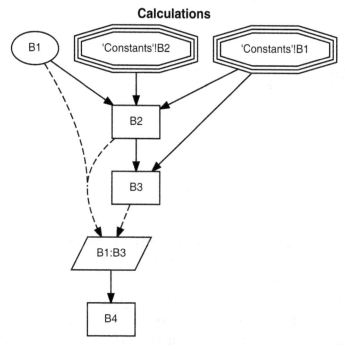

Fig. 11.6. The cells network map redrawn for the Calculations worksheet but without showing the cell contents.

As mentioned earlier, the "Trace Selected Cells" option (Fig. A.3 and A.5) draws the cells network map for only the currently selected cells. For instance, select just cell B2 in the Calculations worksheet, then choose the "Trace Selected Cells" from BuildIt's menu. We will get a chance to specify the name of the picture file and the location in the computer to which to save this file. We will also get an option to determine if we wish to output the cell contents, after which tracing will begin. The cells network map for just cell B2 is as shown in Fig. 11.7.

Calculations

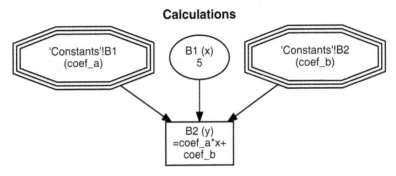

Fig. 11.7. The cells network map for just cell B2 in the Calculations worksheet.

Note that "Trace Selected Cells" traces up to only the first-level precedents of the selected cells. For instance, say cell C10 contains the value 10 and cells C11 and C12 the formulas "=C10" and "=C11", respectively. Tracing the single cell C12 includes itself and its first-level cell precedent C11 in the map but excludes cell C10. Since cell C10 is the second-level precedent of cell C12, cell C10 is not included in the map by the "Trace Selected Cells" option. The "Trace Workbook" option, as we have seen, has no such limitations: all precedent and dependent levels are traced.

11.2 Customizing the cells network map

BuildIt's Trace feature works by programmatically reading and translating the various cell relationships into a DOT-language (Koutsofios and North, 1992) script which is then interpreted by Graphviz (Gansner and North, 2000) to render the cells network map.

As stated earlier, the cells network map is saved as a picture file (GIF format by default). In addition, to this picture file, a plain text file (TXT format) containing the

DOT-language script is also saved. It is this script file that is read by Graphviz and translated into the picture file of the cells network map.

The DOT language is beyond the scope of this book, but you are referred to Graphviz's website at:

www.graphviz.org

for information about the DOT language and Graphviz's features. If you understand the DOT language, you may modify the script file to make the cells network map appear more to your liking.

Nonetheless, some minor customizations are possible even without us having knowledge about the DOT language or understanding how Graphviz works.

BuildIt's Trace keeps a preference file named "ini.txt". This file stores information about the font sizes and various shapes to use by Trace in rendering a cells network map. This preference file is a plain text file which you can open with any text editor and manually modify the preference entries.

Open the location where you stored BuildIt's files (such as C:\BuildIt\). You should see "ini.txt" file together with BuildIt's add-in file "BuildIt.xlam" (for Excel 2007 and onwards) or "BuildIt.xla" (for Excel 2003). If you do not know where BuildIt's files are stored, use your computer's search files features to locate either "BuildIt.xlam" or "BuildIt.xla" file. Once the folder location this add-in file is located, go to that location and you should see the "ini.txt" file in the same location as the BuildIt's add-in file.

However, if you still cannot find the "ini.txt" file (even though you have located the BuildIt's add-in file), you need BuildIt to generate this preference file. Open any Excel workbook (ensure the workbook has some cell entries) and choose "Trace Workbook" from BuildIt's menu. After tracing, the preference file "ini.txt" should appear in the same location of the BuildIt's add-in file.

You should now double click the "ini.txt" file to launch your computer's default file editor (such as Notepad). The preference file contains the following entries with their default values:

```
FONT_TITLE, Helvetica-Bold
FONT_NODE, Helvetica
FONT_TITLE_SIZE, 12
FONT_NODE_SIZE, 10
SHAPE_VALUE, ellipse
SHAPE_FORMULA, box
SHAPE_EMPTY, doublecircle
SHAPE_RANGE, parallelogram
SHAPE_EXTERNAL, tripleoctagon
STYLE_EDGE_RANGE, dashed
OUTPUTFORMAT, gif
```

where FONT_TITLE and FONT_TITLE_SIZE denote the font name and the size of the font (in points), respectively, for the map title. Likewise, FONT_NODE and FONT_NODE_SIZE denote the font name and its size (in points), respectively, for the text inside the shapes. The cell address (and its cell name in brackets, if any) and the cell contents will be displayed using this font and font size.

By default, Traces uses the 12-point Helvetica (bold) font for the map title and the 10-point Helvetica (normal) font for the shape text. The choices of fonts available depends on the fonts that are stored in your computer. But by default, Trace via Graphviz supports the following fonts as listed in Table 11.1.

For instance, to have the Courier font (in bold and in 14 points) for the map title and the Times font (in italics and in 12 points) for the shape text, you need to modify the following preference entries:

```
FONT_TITLE, Courier-Bold
```

```
FONT_NODE, Times-Italic
FONT_TITLE_SIZE, 14
FONT_NODE_SIZE, 12
```

Table 11.1. Font types available for Trace.

AvantGarde-Book	Bookman-Demi
AvantGarde-BookOblique	Bookman-DemiItalic
AvantGarde-Demi	Bookman-Light
AvantGarde-DemiOblique	Bookman-LightItalic
Courier	Helvetica
Courier-Bold	Helvetica-Bold
Courier-BoldOblique	Helvetica-BoldOblique
Courier-Oblique	
Helvetica-Narrow	Palatino-Bold
Helvetica-Narrow-Bold	Palatino-BoldItalic
Helvetica-Narrow-BoldOblique	Palatino-Italic
Helvetica-Narrow-Oblique	Palatino-Roman
Helvetica-Oblique	
NewCenturySchlbk-Bold	Symbol
NewCenturySchlbk-BoldItalic	
NewCenturySchlbk-Italic	
NewCenturySchlbk-Roman	
Times-Bold	ZapfDingbats
Times-BoldItalic	
Times-Italic	
Times-Roman	
ZapfChancery-MediumItalic	

The preference entry SHAPE_FORMULA denotes the shape to use for cells containing formulas, SHAPE_EMPTY for worksheets without any cells interrelationships, SHAPE_RANGE for cell ranges, and SHAPE_EXTERNAL for external cells.

Graphviz supplies nearly 70 types of shapes. Please consult the Graphviz documentation for these shapes.

Nonetheless, Table 11.2 lists some of the available shapes that we might find more applicable in our work.

Table 11.2. Among the shapes available for Trace.

box	octagon	trapezium
circle	parallelogram	triangle
diamond	pentagon	doublecircle
ellipse	polygon	doubleoctagon
hexagon	square	tripleoctagon

Preference entries like the following:

```
SHAPE_FORMULA, diamond
SHAPE_EMPTY, doubleoctagon
SHAPE_RANGE, polygon
SHAPE_EXTERNAL, hexagon
```

instruct Trace to use the diamond shape to represent cells that contain formulas, double-layered octagon for worksheets that have no cells interrelationships, polygon for cell ranges, and hexagon for external cells.

The preference entry STYLE_EDGE_RANGE denotes the type of arrow line style to use for cell ranges (*see* the dashed arrow lines leading to the parallelogram in Fig. 11.1).

For arrow styles, Graphviz supports four styles as shown in Table 11.3.

Table 11.3. Arrow styles available for Trace.

bold	dotted
dashed	solid

The bold style is the same as the solid style but with a thicker line. The dotted style is the same as the dashed style but with smaller, sparser dashes.

Lastly, the preference entry OUTPUTFORMAT denotes the type of picture format to produce for the cells network map. The default is GIF format, but you have over 30 formats to choose (again, please consult Graphviz documentation for the full list). However, the picture formats as listed in Table 11.4 are those we would probably find more applicable in our work.

Table 11.4. Among the output formats available for Trace.

Output	OUTPUTFORMAT value
Windows Bitmap	bmp
Graphics Interchange Format	gif
Joint Photographic Experts Group	jpg, jpeg, or jpe
Macintosh Picture File	pict
Portable Document Format	pdf
Portable Network Graphics	png
Scalable Vector Graphics	svg
Tag Image File Format	tiff

Consequently, the following preference entry:

```
OUTPUTFORMAT, tiff
```

produces the cells network map as a TIFF picture format.

Once the preference file has been modified, save the "ini.txt" file. Please note that changing the preferences in the "ini.txt" file will only affect future cell maps (yet to be generated), not the exisiting maps you had earlier generated using Trace.

11.3 Cells network map for the gcg model

The crop growth and yield model (gcg), as stated earlier, consists of five model components (Fig. 6.1) which we had implemented in an Excel workbook with nine worksheets, which are: Serdang worksheet (holds the weather data), Tables (holds the tabulated data), Control (holds information for BuildIt about the loop, operation, prerun, and option sections), Output (holds the model output), and the five worksheets: Meteorology (implements the Meteorology model component), Photosynthesis (Photosynthesis component), ET (Energy balance component), Water (Soil water component), and Growth (Growth development component).

We can use Trace to produce a cells network map for the gcg's five model components. However, each of these model components consists of a large number of variables and formulas, so a large cells network map will be produced. Consequently, it is better not to combine these individual maps into one map, for such a combined map will be very large and intricate, becoming impractical for us to examine.

Unfortunately, due to page space constraints of this book, it is not possible to show here the cells network maps for the gcg model components in a legible manner. This is because each of these maps is too large to fit within a single page (or even within two pages).

Not all worksheets in the gcg workbook should have their cells interrelationships depicted visually by Trace. Worksheets like Serdang and Tables contain only weather data and tabulated data, respectively; Output worksheet the model output; and Control worksheet the information for BuildIt. Their cells interrelationships are usually of little interest.

A large cells network map can printed using a poster printer, or we can use a poster printing software that is able to split a large image (like a map in our case) into several sections (tiles) and print each of those sections on a A4-sized paper, after which these printed sections are glued together to form the whole image.

11.4 Exercises

1. Use Trace to draw the cells network map for each of the five main gcg model components: Meteorology, Photosynthesis, Energy balance, Soil water, and Crop growth development. Based only on a visual inspection, which of the five gcg model components appears to be the most complex (*e.g.*, the component having the most interconnections between cells) and the least complex? Discuss the implications when a model component has a large number of cell interrelationships.

2. Use Trace to draw the cells network map for the prey-predator model (section 4.3) and the simple plant growth model (section 4.4).

3. Discuss the key strengths and limitations of Trace. For instance, is using Trace to understand a model more useful for small, simple models or for large, complex models?

References

Acock, R. and Reddy, V.R. 1997. Designing an object-oriented structure for crop models. Ecological Modelling, 94: 33-44.

Brown, A.M. 1999. A methodology for simulating biological systems using Microsoft Excel. Computers Methods and Programs in Biomedicine, 58: 181-190.

Caldwell, R.M. and Fernandez, A.A.J. 1998. A generic model of hierarchy for systems analysis and simulation. Agricultural Systems, 57: 197-225.

Campbell, G.S. 1994. Soil Physics with Basic Transport Models for Soil – Plant Systems. Developments in Soil Science 14. 3rd impression. Elsevier Science B.V., Amsterdam, The Netherlands.

Campbell, G.S. and Norman, J.M. 1998. An Introduction to Environmental Biophysics. 2nd Edition. Springer-Verlag, New York.

Choudhury, B.J. and Monteith, J.L. 1988. A four-layer model for the heat budget of homogeneous land surfaces. Quarterly Journal of the Royal Meteorological Society, 114: 373-398.

Dalton, S. 2005. Excel Add-in Development in C/C++. John Wiley & Sons, Ltd., West Sussex, England.

de Jong, J.B.R.M. 1980. Een karakterisering van de zonnestraling in Nederland. Doctor-aalverslag Vakgroep Fysische Aspecten van de Gebouwde Omgeving afd. Bouwkunde en Vakgroep Warmte- en Stromingstechnieken afd. Werktuigbouwkunde, Technische Hogeschool (Techn. Univ.). Eindhoven, Netherlands.

References

Ehleringer, J.R. and Björkman, O. 1977. Quantum yields for CO_2 uptake in C3 and C4 plants: dependence on temperature, CO_2, and O_2 concentration. Plant Physiology, 59: 86-90.

Ephrath, J.E., Goudriaan, J. and Marani, A. 1996. Modelling diurnal patterns of air temperatures, radiation, wind speed and relative humidity by equations for daily characteristics. Agriculture Systems, 51: 377-393.

Farahani, H.J. and Ahuja, L.R. 1996. Evapotranspiration modeling of partial canopy/residue-covered fields. Transactions of the ASAE, 39: 2051-2064.

Frere, M. and Popov, G.F. 1979. Agrometeorological Crop Monitoring and Forecasting. Food and Agriculture Organization of the United Nations, Rome.

Gansner, E.R. and North, S.C. 2000. An open graph visualization system and its applications to software engineering. Software: Practice and Experience, 30:1203-1233.

Goudriaan, J. 1977. Crop Micrometeorology: A Simulation Study. Simulation Monograph. Pudoc, Wageningen.

Goudriaan, J. and van Laar, H.H. 1994. Modeling Potential Crop Growth Processes. A Textbook with Exercise. Current issues in production ecology. Kluwer Academic, Netherlands.

Hansen, F.V. 1993. Surface Roughness Lengths. ARL Technical Report, U.S. Army, White Sands Missile Range, NM 88002-5501.

References

Hillel, D. 1977. Computer Simulation of Soil-Water Dynamics. A Compendium of Recent Work. International Development Research Centre, Ottawa.

Hillyer, C., Bolte, J., van Evert, F. and Lamaker, A., 2003. The MODCOM modular simulation system. European Journal of Agronomy, 18: 333-343.

Hoc, J. 1989. Do we really have conditional statements in our brains? In: E. Soloway and J. Spohrer (Eds.) Studying the novice programmer. Lawrence Earlbaum Associates, Hillsdale, NJ. pp. 179-190.

Kaplanis, S.N. 2005. New methodologies to estimate the hourly global solar radiation; comparisons with existing models. Renewable Energy, 31: 781-790.

Khandan, N.N. 2001. Modeling Tools for Environmental Engineers and Scientists. CRC Press, Boca Raton, Florida.

Koutsofios, E. and North, S. 1992. Drawing Graphs with Dot. Technical report. AT&T Bell Laboratories, Murray Hill, NJ.

Kropff, M.J. 1993. Mechanisms of competition for light. In: M.J. Kropff and H.H. van Laar (Eds). Modelling crop-weed interactions. CAB International (in association with International Rice Research Institute), Wallingford, UK, pp. 33-61.

Kruck, S.E. 2006. Testing spreadsheet accuracy theory. Information and Software Technology, 48: 204-213.

Kustas, W.P. and Norman, J.M. 1999. Evaluation of soil and vegetation heat flux predictions using a simple two-source model with radiometric temperatures for partial canopy cover. Agricultural and Forest Meteorology, 94: 13-29.

Lafleur, P.M. and Rose, W.R., 1990. Application of an energy combination model for evaporation from sparse canopies. Agricultural and Forest Meteorology, 49, 135-153.

Lewis, C. and Olson, G. 1987. Can principles of cognition lower the barriers to programming? In: G.M. Olson, S. Sheppard and E. Soloway (Eds.) Empirical Studies of Programmers: Second Workshop. Norwood, NJ: Ablex. pp. 248-263.

Liebowitz, S.E. and Margolis, S.J. 1999. Winners, Losers & Microsoft: Competition and Antitrust in High Technology. The Independent Institute, Oakland, California.

Mitchell, J.W. 1976. Heat transfers from spheres and other animal forms. Biophysical Journal, 16: 561-569.

Miyazaki, T. 2005. Water Flow in Soils. CRC Press, Boca Raton, FL.

Nardi, B. and Miller, J.R. 1990. The spreadsheet interface: a basis for end user programming. INTERACT'90, Elsevier Science Publishers B.V., North Holland, pp. 977-983.

Ortega-Farias, S., Acevedo, C. and Fuentes, S. 2000. Calibration of the Penman-Monteith method to estimate latent heat flux over a grass canopy. Acta Horticulturae, 475: 129–133.

Papajorgji, P., Beck, W.B. and Braga, J.L. 2004. An architecture for developing service oriented and component-based environmental models. Ecological Modelling, 179: 61-76

Papajorgji, P.J. and Pardalos, P.M. 2006. Software engineering techniques applied to agricultural systems: an object-oriented and UML approach. Springer Science + Business Media Inc., New York.

Raffensperger, J.F. 2003. New guidelines for spreadsheets, International Journal of Business and Economics, 2: 141-154.

Rajalingham, K., Chadwick, D.R. and Knight, B. 2001. Classification of spreadsheet errors. In: Symposium of the European Spreadsheet Risks Interest Group (EuSpRIG), pp. 23-34.

Read, N. and Batson, J. 1999. Spreadsheet Modelling Good Practice. IBM & Institute of Chartered Accountants in England and Wales, London.

Scoville, R. 1994. Spreadsheets. The PC Bible edition. E. Knorr. Peachpit Press, California.

Seila, A.F. 2005. Spreadsheet simulation. In: M.E. Kuhl, N.M. Steiger, F.B. Armstrong, and J.A. Joines (Eds.) Proceedings of the 2005 Winter Simulation Conference, Orlando, Florida, pp. 33-40.

Shuttleworth, W.J. and Wallace, J.S. 1985. Evaporation from sparse crops – an energy combination theory. Quarterly Journal of the Royal Meteorological Society, 111: 839-855.

Soloway, E., Bonar, J. and Ehrlich, K. 1983. Cognitive strategies and looping constructs: an empirical study. Communications of the ACM, 26: 853-860.

Spitters, C.J.T., 1989. Weeds: population dynamics, germination and competition. In: R. Rabbinge, S.A. Ward, H.H. van Laar (Eds.), Simulation and Systems Management in Crop Protection. Pudoc, Wageningen, pp. 217-239.

Stannard, D.I. 1993. Comparison of Penman-Monteith, Shuttleworth-Wallace, and modified Priestley-Taylor evapotranspiration models for wildland vegetation in semiarid rangeland. Water Resources Research, 29: 1379–1392.

Syrstad, T. and Jelen, B. 2004. VBA and Macros for Microsoft Excel. Sams Publishing, Indianapolis, Indiana.

Szeicz, G. and Long, I.F., 1969. Surface resistance of crop canopies. Water Resources Research, 5: 622-633.

Teh, C.B.S., Henson, I.E., Goh, K.J. and Husni, M.H.A. 2004. The effect of leaf shape on solar radiation interception. In: Hawa Z.E. Jaafar et al. (Eds). Proceedings of the 15th. Malaysian Society of Plant Physiology Conference. 14-16 September 2004, Port Dickson, Negeri Sembilan. pp. 48-58

Teh, C.B.S. 2006. Introduction to Mathematical Modeling of Crop Growth: How the Equations are Derived and Assembled into a Computer Program. BrownWalker Press, Boca Raton, Florida.

Teh, C.B.S. 2011. Overcoming Microsoft Excel's weaknesses for crop model building and simulations. Journal of Natural Resources and Life Sciences Education, 40: 122-136.

Thornley, J.H.M. 1976. Mathematical Models in Plant Physiology. Academic Press, London.

van Keulen, H. and Seligman, N.G. 1987. Simulation of Water Use, Nitrogen and Growth of a Spring Wheat Crop. Simulation Monographs. Pudoc, Wageningen.

van Kraalingen, D.W.G., Rappoldt, C. and van Laar, H.H. 2003. The Fortran simulation translator, a simulation language. European Journal of Agronomy, 18: 359-361.

Yin, X. and van Laar, H.H. 2005. Crop Systems Dynamics. An Ecophysiological Simulation Model for Genotype-by-environment Intreactions. Wageningen Academic Publishers, The Netherlands.

Appendix A. Installation of BuildIt

Pre-requisites

BuildIt is an add-in that works only in Microsoft Excel version 2003 onwards (including version 2007, 2010, and 2013). BuildIt was not tested on lower versions than Excel 2003. Additionally, BuildIt works only in the Microsoft Windows operating system.

BuildIt's Trace features requires the Graphviz software from AT&T Research Lab to work. Graphviz has to be downloaded separately. This is to ensure you will obtain the latest version of Graphviz, which may have speed improvements, enhancements, and bug fixes.

The following are the steps to install BuildIt:
1. Download and unpack the BuildIt ZIP archive
2. Load BuildIt into Excel
3. Download and install Graphviz
4. Set the computer's PATH variable to include Graphviz's path

1. Download and unpack the BuildIt ZIP archive

BuildIt is free and can be downloaded from:

www.christopherteh.com/buildit

Once downloaded, use Windows Explorer to find and double click the "buildit.zip" file. Downloaded files are typically stored in one common or default location in your computer (such as the "Downloads" folder or on your computer desktop). Your computer's default file archive program will open when you double click the archive file.

In the "buildit.zip" archive, you will notice several folders:

1. "Excel 2003" folder stores BuildIt add-in for Excel 2003,
2. "Excel 2007 and above" folder stores the BuildIt add-in for Excel 2007 and onwards,
3. "Book examples" folder stores the worksheets containing the examples used in this book,
4. "Weather" folder contains two worksheets: one for the Serdang weather and the other for Netherlands, and
5. "Install" folder contains files for displaying help on installing BuildIt,

Depending on what version of Excel you are using, extract the required BuildIt add-in. Ensure you use the correct BuildIt add-in for the Excel version you are using.

You will also need to extract the Serdang weather worksheet (optional: Netherlands weather worksheet). You can optionally extract the book example worksheets.

Unpack the archive contents to any location you wish in your computer. But it is recommend that you save BuildIt files in its own folder, *e.g.*, "C:\BuildIt\".

For the next step, follow step 2a if you are using Excel 2007 or a higher version, else step 2b if you are using Excel 2003.

2a. Load BuildIt add-in into Excel 2007 and onwards

Launch to open Excel. In Excel 2010 and 2013, click the "File" menu and choose "Options". Excel 2007 has a slightly different layout. In Excel 2007, click the "Office" icon at the upper left corner of Excel's window, and choose "Options".

From the Options dialog box, choose the "Add-Ins", then in the "Manage:" dropdown list, select "Excel Add-ins", and click the "Go..." button (Fig. A.1).

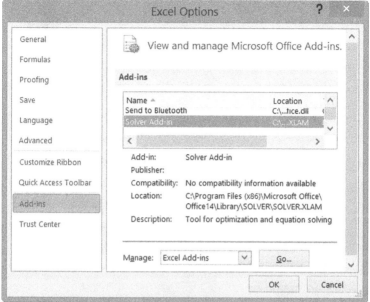

Fig. A.1. Excel Options dialog box to manage Excel add-ins.

You will be presented with a list of all the add-ins recognized by Excel (Fig. A.2). If you had saved BuildIt's files in the default add-in location, you will see the name BuildIt on this list.

However, if you do not see BuildIt on this list, this only means that you had saved BuildIt in a non-default location. In this case, click the "Browse..." button to locate BuildIt's installation folder (*e.g.*, "C:\BuildIt\").

Once you have found BuildIt's location, you will see BuildIt's name appear on the list (Fig. A.2). Ensure that you place a tick mark next to BuildIt's name. Placing a tick mark next to BuildIt's name will instruct Excel to load the BuildIt add-in.

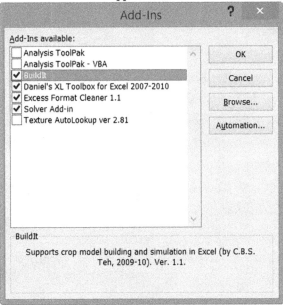

Fig. A.2. List of add-ins recognized by Excel. This list will appear differently on your computer, depending on what add-ins you have installed.

Click the "OK" button, and once loaded, BuildIt's menu will appear on the Ribbon (Fig. A.3). BuildIt is now ready to use each time you open Excel.

Fig. A.3. BuildIt's menu added onto Excel's Ribbon.

2b. Load BuildIt add-in into Excel 2003

On Excel's main menu, click "Tools" and then "Add-Ins..." (Fig. A.4).

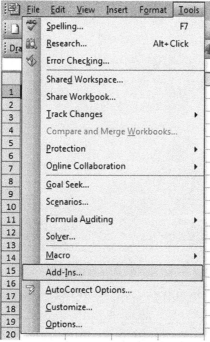

Fig. A.4. Managing the add-ins in Excel 2003.

You will see a list of add-ins recognized by Excel (same as Fig. A.2). If you had saved BuildIt's files in the default add-in location, you will see the name BuildIt on this list.

However, if you do not see BuildIt on this list, this only means that you had saved BuildIt in a non-default location. Click the "Browse..." button to locate BuildIt's installation folder (*e.g.*, "C:\BuildIt\"). Once you have found BuildIt's location, you will see BuildIt's name appear on the list (Fig. A.2). Ensure that you place a tick mark next to BuildIt's name. Placing a tick mark next to BuildIt's name will instruct Excel to load the BuildIt add-in.

Click the "OK" button, and once installed, BuildIt's menu will appear on the main menu (Fig. A.5). BuildIt is now ready to use each time you open Excel.

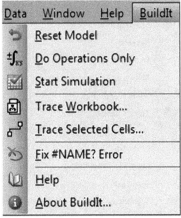

Fig. A.5. BuildIt's menu on Excel 2003's main menu.

2c. *Unload and uninstall BuildIt*

Unloading BuildIt means BuildIt is no longer available for use in Excel. However, BuildIt files are not removed or deleted from your computer. BuildIt files still remain in your computer, but Excel has unloaded (will not use) BuildIt.

To unload BuildIt, remove the tick mark next to BuildIt's name in the add-in list (Fig. A.2). This immediately unloads BuildIt and stops Excel from loading BuildIt each time Excel starts. If you decide to use BuildIt again, load it into Excel by placing a tick mark beside BuildIt's name in the add-in list.

To completely remove BuildIt from your computer, manually delete all of BuildIt's files in its installation folder. If you attempt to load BuildIt, Excel will produce an error message that the BuildIt add-in file cannot be found, and Excel will remove BuildIt's name from the add-in list. Uninstallation is now complete.

3. Installing Graphviz

At this time of writing, Graphviz is currently at version 2.38. You can download for free Graphviz from:

www.graphviz.org/Download_windows.php

Ensure you are downloading the Windows version. Graphviz installation file comes either in MSI or ZIP format. You can download either of the two formats.

Once downloaded, double click the MSI (or ZIP) file to start the Graphviz installation. Follow the given instructions during installation.

4. Set the computer's PATH variable

PATH is a Windows' environment variable that contains the locations of certain executable programs in the computer. Because their locations are stored in the PATH variable, these programs can be launched (run) from anywhere in the computer. This is very convenient because we do not need to run these executable files from their location.

Trace uses Graphviz's "dot.exe" file to render cells maps. Because your Excel workbook can be anywhere in the computer, Trace needs to be able to run "dot.exe" regardless of the current location. This ability is only possible provided the location of Graphviz executables is specified in the PATH variable.

Earlier versions of Graphviz had simpler installations. Graphviz's installer would automatically add the location of Graphviz executable files to the variable PATH. Unfortunately, this automatic path specification is no longer supported in recent versions of Graphviz. This means we have to manually add the location of Graphviz executables to the PATH variable.

By default, Graphviz executable files are installed in:

"C:\Program Files (x86)\Graphviz*VersionNumber*\bin"

or

"C:\Program Files\Graphviz*VersionNumber*\bin".

So, if you are installing Graphviz version 2.38 (the latest version at the moment), the executable files' location would either be:

"C:\Program Files (x86)\Graphviz2.38\bin"

or

"C:\Program Files\Graphviz2.38\bin".

Use Windows Explorer to confirm the exact location for Graphviz executables (*e.g.*, check where "dot.exe" file is stored by using the Windows Explorer file search).

To add Graphviz executables' path to the PATH variable, do the following.

As a shortcut, press the Windows logo ⊞ key (also called the Windows key or Start key) together with the letter "r" key to launch the Run dialog box (Fig. A.6).

In the Run dialog box, type the following command:

```
cmd
```

at the "Open:" prompt, which would then open the command-line prompt (Fig. A.7).

At the command prompt, type the following:

```
setx path "%path%;C:\Program Files\Graphviz2.38\bin"
```

which includes the location of Graphviz executables in the PATH variable. In the above case, Graphviz's location is "C:\Program Files\Graphviz2.38\bin".

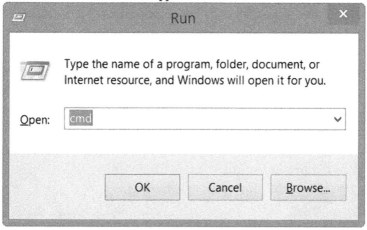

Fig. A.6. Run dialog box. The cmd command is to open the command-line prompt.

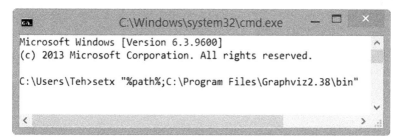

Fig. A.7. Command-line prompt. The setx command is used to set the PATH variable to include the location of Graphviz executable files.

Make the necessary change if this location is not the same as for yours, *e.g.*,

```
setx path "%path%;C:\Graphviz2.38\bin"
```

where in this case the location of the Graphviz executables is in "C:\Graphviz2.38\bin".

In both these cases, the `setx` command is used to set the PATH variable. Note the use of the quotation marks and the single semicolon immediately after the `%path%` text.

Close the command-line prompt, and restart Excel for the changes to take effect. Depending on your computer system, you may need to reboot the computer before these changes take effect.

If the setup is correct, you should be able to use the Trace feature in BuildIt.

Appendix B. BuildIt menu commands

From Fig. A.3, the BuildIt's menu in Excel 2007 and above is:

while from Fig. A.5, the BuildIt's menu in Excel 2003 is:

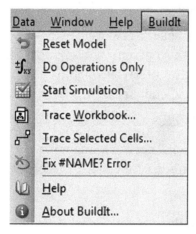

Start Simulation

This command is used when model simulations require a loop for repetitious calculations. Consequently, this command requires cell names _step, _stepsize, and _criteria to be defined first in the current workbook.

You can stop a simulation run at any time by pressing the Esc key or the Control-C keys.

Do Operations Only

Executes all actions listed only in the operation section, without starting the loop, if any.

Reset Model

Executes all actions listed only in the prerun section, without starting the loop, if any. This command is often used when you have INI actions to initialize variables. Using this command will then give these variables their initial values.

Trace Workbook

Produces a cells network map for one or more worksheets in the current workbook. This map depicts visually all cells interrelationships. All levels of cells precedents and dependents are traced.

Trace Selected Cells

Produces a cells network map for only the currently selected cells. Only first-level cell precedents are traced.

Fix #NAME? Error

This command fixes the Excel's error message that one or more external links in a workbook cannot be updated. The exact error message, which depends on your Excel version, would resemble the following:

"This workbook contains one or more links that cannot be updated"

This error happens when a workbook uses the BuildIt's functions (interpolate or solve), but BuildIt was not installed in the same location in all your computers. Consequently, Excel is unable to locate BuildIt in the computer you are currently using to open your workbook.

Because of this update link error, cells that use the interpolate or solve functions will return the #NAME? error value.

To fix this error, just choose the "Fix #NAME? Error" command from BuildIt's menu each time you open the workbook in different computers.

Note this command works to fix only BuildIt's links and not the links to other sources, add-ins, or programs.

Help

Displays a brief help file.

About BuildIt

Displays information about the BuildIt add-in.

Index